Laboratory Manual in Food Chemistry

other AVI books

BASIC FOOD CHEMISTRY Cloth and Soft Cover *Lee*
CARBOHYDRATES AND THEIR ROLES *Schultz, Cain and Wrolstad*
DAIRY LIPIDS AND LIPID METABOLISM *Brink and Kritchevsky*
DAIRY TECHNOLOGY AND ENGINEERING Cloth and Soft Cover *Harper and Hall*
ELEMENTARY FOOD SCIENCE Cloth and Soft Cover *Nickerson and Ronsivalli*
ENCYCLOPEDIA OF FOOD ENGINEERING *Hall, Farrall and Rippen*
ENCYCLOPEDIA OF FOOD TECHNOLOGY *Johnson and Peterson*
FOOD ANALYSIS LABORATORY EXPERIMENTS Soft Cover *Meloan and Pomeranz*
FOOD ANALYSIS: THEORY AND PRACTICE *Pomeranz and Meloan*
FOOD CHEMISTRY *Aurand and Woods*
FOOD CHEMISTRY Soft Cover *Meyer*
FOOD COLORIMETRY: THEORY AND APPLICATIONS *Francis and Clydesdale*
FOOD DEHYDRATION, 2ND EDITION, VOLS. 1 AND 2 *Van Arsdel, Copley and Morgan*
FOOD ENGINEERING SYSTEMS, VOL. 1 OPERATIONS Cloth and Soft Cover *Farrall*
FOOD ENZYMES *Schultz*
FOOD PROCESS ENGINEERING Cloth and Soft Cover *Heldman*
FOOD SCIENCE, 2ND EDITION *Potter*
FUNDAMENTALS OF DAIRY CHEMISTRY, 2ND EDITION *Webb, Johnson and Alford*
FUNDAMENTALS OF FOOD ENGINEERING, 2ND EDITION *Charm*
LIPIDS AND THEIR OXIDATION *Schultz, Day and Sinnhuber*
MICROBIOLOGY OF FOOD FERMENTATIONS *Pederson*
MILESTONES IN NUTRITION *Goldblith and Joslyn*
NUTRITIONAL EVALUATION OF FOOD PROCESSING, 2ND EDITION Cloth and Soft Cover *Harris and Karmas*
PRACTICAL FOOD MICROBIOLOGY AND TECHNOLOGY, 2ND EDITION *Weiser, Mountney and Gould*
PROBIOTICS *Sperti*
PROGRESS IN HUMAN NUTRITION, VOL. 1 *Margen*
PROTEINS AND THEIR REACTIONS *Schultz and Anglemier*
PROTEINS AS HUMAN FOOD American Edition *Lawrie*
INTERNATIONAL SYMPOSIUM: SELENIUM IN BIOMEDICINE *Muth, Oldfield and Weswig*
SYMPOSIUM: PHOSPHATES IN FOOD PROCESSING *deMan and Melnychyn*
SYMPOSIUM: SULFUR IN NUTRITION *Muth and Oldfield*
THE CHEMISTRY AND PHYSIOLOGY OF FLAVORS *Schultz, Day and Libbey*
THE SAFETY OF FOODS *Graham*
THE TECHNOLOGY OF FOOD PRESERVATION, 3RD EDITION *Desrosier*

Laboratory Manual in Food Chemistry

by A. E. Woods, Ph.D.
Professor of Chemistry,
Middle Tennessee State University,
Murfreesboro, Tennessee

and L. W. Aurand, Ph.D.
Professor of Food Science and
Biochemistry, North Carolina
State University,
Raleigh, North Carolina

THE AVI PUBLISHING COMPANY, INC.
WESTPORT, CONNECTICUT

© *Copyright 1977 by*

THE AVI PUBLISHING COMPANY, INC.

Westport, Connecticut

All rights reserved. No part of this work covered by the copyright hereon may be produced or used in any form or by any means—graphic, electronic, or mechanical, including photocopying, recording, taping, or information storage and retrieval systems—without written permission of the publisher.

ISBN-0-87055-220-1

Printed in the United States of America

Preface

This book was designed entirely with the needs of the student in mind. The methods and analytical techniques were chosen for their reliability and use of newer aspects of food analysis. Within certain limits, the majority of the laboratory experiments can be performed during a period of three hours. A large number of the experiments are classroom-tested by our students for which we are grateful.

It was the intent of the authors to design the text for use as a supplement to their textbook *Food Chemistry*. Consequently, no attempt was made to include long descriptions of the theory; rather it was our intent to include sufficient laboratory work for the student to develop an understanding of modern food chemistry.

The authors wish to acknowledge Dr. Donald Tressler, Mr. T. L. Amerson, Jr., Mr. James Taylor and our fellow workers and colleagues. Appreciation is extended to our wives Saundra and Eleanor.

Special acknowledgements are in order for Parr Instrument Company, Roche Chemical Division, Hoffmann-LaRoche Incorporated; Pierce Chemical Company; and Perkin-Elmer Corporation.

A. E. WOODS
L. W. AURAND

November 1976

Contents

Introduction

Rapid progress has been made in experimental research because suitable analytical procedures have been made available to determine either the composition of substances or the progress of reactions. The object, therefore, of experimental research is to ascertain what substance(s) is present in a material (qualitative analysis) and/or how much of the substance(s) is present in the material (quantitative analysis). It is at once apparent that in any investigation a qualitative analysis must precede a quantitative analysis so that the appropriate analytical procedures may be used. In food analysis, the various components of a food are usually well known; hence it is simply a matter of selecting the proper procedure for the component in question.

In the successful execution of any analysis, it is essential that the principles underlying the procedure be understood before attempting the actual determination. Unless this is done, the analyst follows a "cook book" procedure from which he gains little or no knowledge and runs into the possibility of serious error due to lack of understanding of the determination. In addition to understanding the procedures, the student should be familiar with some of the problems encountered in the gathering of data.

THE NOTE BOOK

A chemist's notebook should be a complete record of all work performed. All data must be recorded in permanent form as soon as they are obtained. The entries should be made in ink. No erasing is permitted, and any necessary corrections should be made without destroying the original entries.

The notebook should show dates when the analyses were started and completed; it should contain the sample number or sample description for each determination; it should contain sample sizes, volumes, aliquots, titration values, readings of instruments, etc., and the results and conclusions of the analysis should be recorded.

All determinations should be performed in duplicate, for in spite of the care taken by an analyst to avoid errors, mistakes will occur during analysis. For example, an error in the reading of an instrument or the inadvertent loss of a material during a transfer step may pass without detection if samples are not run in duplicate.

SIGNIFICANT NUMBERS

Considerable loss of time and effort may occur when an analyst does not appreciate the fact that the accuracy of calculated results is dependent on the accuracy of the procedure. It is unproductive to record and calculate analytical data with more or less significant figures than are justified by the accuracy of the experiment.

A significant figure is considered to include all the digits with definitely known values and the first indefinite digit. Ciphers may be significant digits or may be used to locate the position of a decimal point. In the quantity 10.23 there are four significant figures whereas in the quantity 0.023 there are two significant numbers.

A general rule for recording analytical data is to retain only one uncertain digit. For example, in weighing a sample to the nearest milligram the weight is recorded to the third decimal point. It is incorrect to add another digit which would report the weight to four decimal places.

The number of significant figures to keep in the calculation of results is very important. In adding or subtracting, and in the sum or difference, keep in each number only as many digits to the right of the decimal as there are to the first indefinite digit. For example, in adding the following terms: $1.23 + 2.3456 + 3.451$, in which it is assumed that the last digit of each term is indefinite the sum is $1.23 + 2.35 + 3.45 = 7.03$. Similarly, in multiplication and division, keep as many digits in each value and in the results which will represent an uncertainty no greater than the quantity having the fewest significant digits. For example, in calculating the value of

$$\frac{11.22 \times 0.0134}{2.5575}$$

the figures should be rounded off to:

$$\frac{11.2 \times 0.0134}{2.56} = 0.0586$$

Note: when extra figures are dropped from a quantity, interpret the value of the last significant digit as ±; i.e., increase the last digit by 1 if the following rejected digit(s) is over 5 and vice versa. There are situations in which the retention of significant figures is a matter of individual judgment. For example, if one were to apply the above rules to computations in analysis; the product of 12.3 × 0.0045 × 1.678 would be written 12.3 × 0.0045 × 1.7. The second factor 0.0045 has only two significant figures and hence its value is stated with an implied uncertainty of 1 part in 45 or roughly 2%. However, if we reduce the other two factors to two significant factors, the latter factor is expressed as 1.7 with an implied uncertainty of 1 part in 17 or roughly 6%. Therefore, the first and third factors should contain 3 significant figures in order to express an uncertainty no greater than 2% (0.0045).

ERRORS

No analytical procedure is entirely free from possible error and, as a consequence, the accuracy of the data is dependent on the reduction of that error. The magnitude of the error involved in any individual analytical method is dependent on several factors. These factors include:

(1) Precision of the individual analyst.
(2) Accuracy of the instrument or instruments used.
(3) Kind of analytical method employed.
(4) Quantity of substance analyzed.

As noted above, all analyses are done in duplicate and, in general, it is assumed that if the results agree with each other closely, they are correct. This is usually true but it happens occasionally that though duplicate results may agree they are nevertheless incorrect because errors in results can compensate one another as well as be cumulative. Obviously, if the duplicates fail to check each other it is a good indication that neither result can be considered reliable. If two different methods of analysis are used and the results obtained check very closely they are more likely to be correct.

Intrinsically, there are two kinds of errors (in analytical work) when one measures a physical quantity: random errors and systematic errors. Random errors are those errors which vary during the same experiment. The size of this kind of error is dependent upon the precision of the analyst. They may be positive or negative and they can be analyzed statistically. Systematic errors are those errors which are constant during the same experiment. This kind of error is due to inaccurate instruments or to the design of the experiment. The results may be high or low and they cannot be analyzed statistically.

BRIEF SURVEY OF RANDOM ERRORS IN THE MEASUREMENT OF FOOD CONSTITUENTS

In working with unknown samples one does not know the true values for the samples. However, true value may be approximated by calculating the arithmetic mean ($\bar{x}$) of a number of determinations:

$$\bar{x} = \frac{\Sigma X}{n}$$

where ΣX = sum of all the X's and n = number of readings. The greater the number of readings that one obtains the closer the arithmetic mean comes to the true value, assuming that no systematic errors are present.

The degree to which numerical data tends to be distributed about an average value is called the variation of the data. Various measures of variation are available, the two most common being mean deviation and standard deviation.

The mean deviation of a number of readings is defined by:

$$\text{M.D.} = \Sigma(X - \bar{x})$$

where $\bar{x}$ is the arithmetic mean of the numbers and $(X - \bar{x})$ is the absolute value of the difference between any individual reading and the arithmetic mean.

The dispersion of results about the mean is given in terms of the standard deviation (σ) and is defined by:

$$\sigma = \frac{\sqrt{\Sigma(X-\mu)^2}}{N}$$

where μ is the true mean and N is the population size.

Or

$$s = \frac{\sqrt{\Sigma(X-\bar{x})^2}}{n}$$

when computing from the sample. The standard error of the means is given

$$\sigma_{\bar{x}} = \sigma(\text{of } \bar{x}) = \frac{\sigma}{\sqrt{n}}$$

It is necessary sometimes to differentiate the standard deviation of a population from the standard deviation of a sample taken from this population. The symbol s is used for the latter and σ for the former.

THE NORMAL DISTRIBUTION

If a large number of observations are made for the same physical quantity, the results tend to follow a normal distribution curve (Fig. I.1). The curve can be defined by the equation

$$Y = \frac{1}{\sigma\sqrt{2\pi}} e^{-\frac{1}{2}\frac{(X-\mu)^2}{\sigma}}$$

where μ = mean, σ = standard deviation, π = 3.14159, e = 2.71828. When μ = o, the resulting distribution is often called the "error function" and σ = root mean square error.

The integrated form of the normal curve can be used to compute probabilities that the errors will fall within certain limits. Thus, the integral

$$\int_a^b YdY = \frac{1}{\sigma\sqrt{2\pi}} \int_a^b e^{\frac{1}{2}\left(\frac{X-\mu}{\sigma}\right)^2} dX =$$

Probability $[a<X<b]$

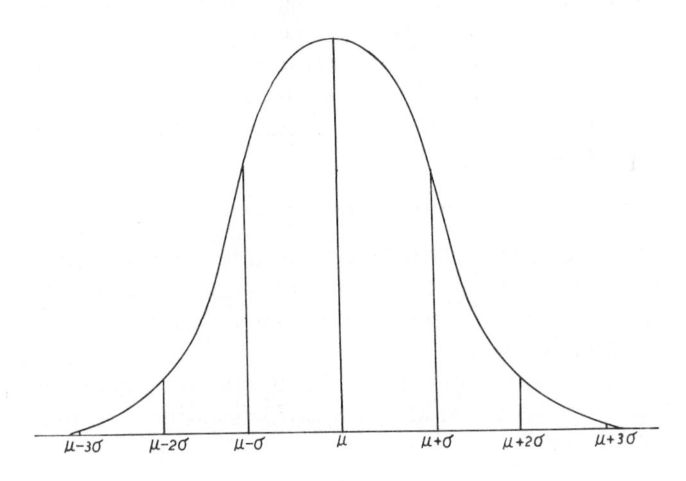

FIG. I.1

A NORMAL CURVE FOR THE DISTRIBUTION OF EXPERIMENTAL RESULTS

We also note that setting $a = -\infty$ and $b = +\infty$.

There are tables of ratios of partial areas to the whole area bounded by the normal curve, corresponding to various values of X/σ, from which the probability of a value falling within or outside a range can be determined. For normal distributions: (a) The range $\mu\pm\sigma$ (i.e., one standard deviation on either side of the mean) includes 68% of the values. Stated another way, there is a P of 0.68 that a value of X falls within this range and a P of 0.32 that a value falls outside this range. (b) The range $\mu\pm2\sigma$ includes 95% of the values (i.e., two standard deviations on either side of mean). There is a P of 0.05 that a value of X falls outside this range. (c) The range $\mu\pm3\sigma$ includes 99% of the values (i.e., three standard deviations on either side of the mean). There is a P of 0.01 that a value of X falls outside this range.

The normal range for biological data is given by $\bar{x}\pm2s$, where s is the standard deviation of the sample and $\bar{x}$ = sample mean.

GRAPHS

A graph is a pictorial presentation showing the change of one variable with the change in another. The type of graph employed will depend on the nature of the data involved and the purpose for which the graph is intended.

In all graphs it is commonly accepted that the independent variable, or X, is plotted along the abscissa ($Y = 0$ line), and the dependent variable, or Y, is plotted along the ordinate ($X = 0$ line). Thus, Y is said to be a function of X, which means that its value is a function of X. Schematically it is written

$$Y = f(X)$$

Several salient features should be considered when plotting graphs.

(1) The scales along the axes should be simple numbers.

(2) The experimental points should be clearly indicated and should not be crowded together.

(3) A smooth curve should be fitted between the experimental points rather than joining each point together. If not, a jagged curve often results implying an unlikely relationship between the variables.

(4) It is unnecessary to begin both ordinate and abscissa at 0 unless the curve represents a straight line satisfying the relationship $Y = mX$. If the graph represents a straight line satisfying the relationship $Y = mX + b$, only the X axis (abscissa) need begin at zero.

GRAPHING

Another function of a graph is to determine the relationship between the two variables ($Y = mX$), and the constants which correlate these two variables ($Y = mX + b$, where m and X are the variables).

A linear function of X has the general form $Y = mX + b$, where m is the slope and b is the intercept. Its graph is a straight line (Fig. I.2). The slope m is the number of units change in Y per unit change in X. The intercept b is the value of Y where the graph intercepts the Y axis ($X = 0$).

In plotting a straight line graph it is common to begin the abscissa (X axis) at zero. If, however, the intercept on the X axis ($-b/m$) is required, it is necessary to begin the Y axis on zero and to plot the X axis in the negative as well as positive direction.

It is possible to plot a function of higher degree by substitution or rearrangement so that a straight line graph is obtained. For example,

(1) The parabolic curve ($Y = kX^2$) can be converted to the linear form. In a relationship of this type we let X^2 be the variable rather than X. It will be noted that when one is plotting a

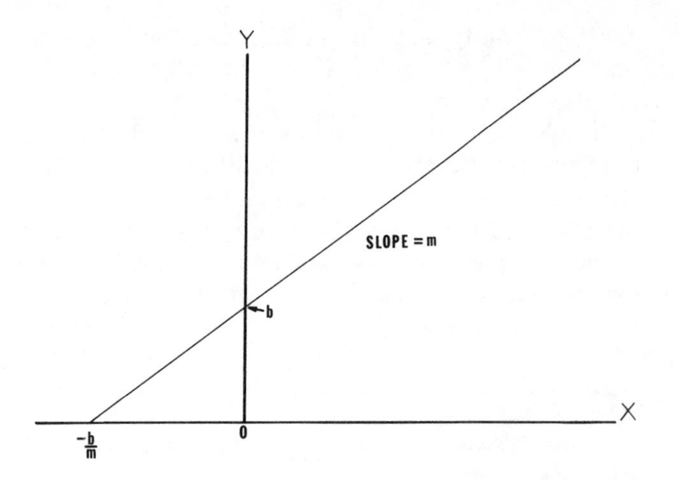

FIG. I.2
GRAPH SATISFYING THE EQUATION Y = mX + b

function of X along the abscissa, the experimental points should be evenly spaced. Similarly, in this case, choose values of X which will give values of X^2 evenly spaced along the abscissa.

(2) The hyperbolic curve can be converted to the linear form by plotting the reciprocals of the variables. An example of this type is the Lineweaver-Burk plot of enzyme kinetics:

$$v = \frac{V\max S}{K_m + S}$$

where v (initial reaction velocity) and S (substrate concentration) are variables.

Rearranging to separate the variables v and S:

$$v(K_m + S) = V\max S$$

Converting to the reciprocal:

$$\frac{1}{v(K_m + S)} = \frac{1}{V\max S}$$

Transpose to give:

$$\frac{1}{v} = \frac{K_m}{V\max S} + \frac{S}{V\max S}$$

Thus:

$$\frac{1}{v} = \frac{K_m}{V\max} \cdot \frac{1}{S} + \frac{1}{V\max} \text{ (The form is } y = mx + b.)$$

(3) The exponential curve ($Y = ab^X$) can be obtained as a straight line by plotting the log of Y versus X. An example of this type is the curve one obtains for growth of bacteria during the logarithmic phase (Fig. I.3).

The above relationship is as follows:

$$N = N_0 \, \epsilon^{\alpha t}$$

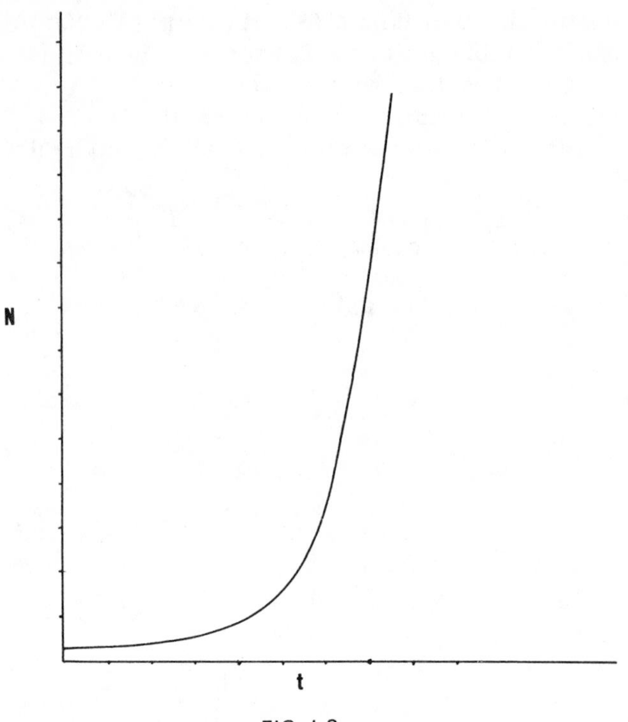

FIG. I.3
GRAPH OF BACTERIA GROWTH DURING LOGARITHMIC PHASE

Where N_0 is number of viable cells at zero time, N = present number of viable cells at time t, and α = a constant.

In converting the equation to log form:

$$\frac{N}{N_0} = \alpha t$$

Converting to base 10 logs:

$$2.303 \log \frac{N}{N_0} = \alpha t = \log_{10} \frac{N}{N_0} = At$$

$$(\text{A} = \text{constant} = 0.4343\alpha)$$

or

$$\log_{10} N = \log_{10} N_0 + At$$

The form is $Y = b + mX$, where b is $\log_{10} N_0$ and m is equal to A.

To facilitate this, we frequently employ special graph paper which has one of the scales (Y axis) calibrated logarithmically. This is referred to as semi-log paper. A straight line is obtained by plotting N (Y) against time (X).

(4) The geometric curve ($Y = a\,X^b$ or $\log Y = \log a + b \log X$) will show a linear relationship by plotting Y against X on log-log graph paper. This type of graph is useful in determining the order of a multimolecular reaction.

To obtain the constants, we graph the data and draw the best straight line through the points.

Given any two points (X_1, Y_1) and (X_2, Y_2) on the lines, the constants m and b can be calculated in the following manner:

(1) Slope $(m) = \frac{Y_2 - X_1}{X_2 - X_1}$

(2) Intercept (b) is value of Y when X = O.

Solutions

PHYSICAL PROPERTIES OF SOLUTIONS

A solution may be defined as a physically and chemically homogeneous mixture of two or more substances in which the solute is in an ionic or molecular state of subdivision and the solvent is the substance (usually a liquid) capable of dissolving the solute.

The solubility of a substance depends upon the tendency of its molecules to go into solution. When a substance is in dilute solution it conforms to the gas law ($PV = nRT$) and it tends to distribute itself uniformly throughout the solvent by diffusion. If a solution consisting of solute and solvent (water) is separated from the solvent *per se* by a membrane permeable to the solvent, the solute will exert an osmotic pressure. Ions, molecules and micelles contribute to osmotic pressure in proportion to the number of kinetically active particles per unit volume. If the solute is nonvolatile, only the solvent can cross the membrane. Osmotic flow is, therefore, the result of vapor pressure difference between solvent and solution. Osmotic pressure is the driving force of diffusion. Osmotic pressure is expressed as $\pi = mRT$ where π = pressure, m = molality, R = gas constant and $T = °K$.

Osmotic Pressure of Solutions and Permeability of Membranes

The preparation of suitable membranes is one of the problems of osmometry. As the size of the particles increases, sieve type membranes such as cellophane can be used. Oxidized nitrocellulose membranes are negatively charged and are impervious to anions. Copper ferrocyanide membranes are semipermeable to sugar solutions.

One problem associated with osmotic pressure measurements is the actual change in the level of the liquid in the osmometer tube (Fig. 1.1). If the level of the liquid is observed in tubes of large diameter, it will take a relatively large volume of osmotic flow to produce a visible change. In contrast, if a capillary tube is used, a large level difference will be noted. The meniscus of water tends to stick in a capillary tube because of imperfect wetting; hence, a liquid of low surface tension and good wetting properties is often used. The meniscus is in an enlargement where surface forces are unimportant.

Procedure.—Set up several osmometers using collodion sacs illustrated in Figure 1.1.

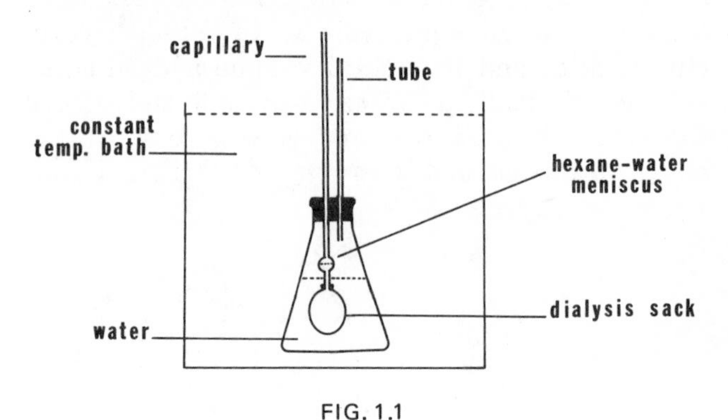

FIG. 1.1
OSMOTIC PRESSURE APPARATUS

(A) Fill one sac with $M/64$ sucrose, and another with $M/128$ sodium chloride. Immerse the sacs into a flask containing distilled water. Solutions and water should be at the same temperature as the bath. Allow to stand until equilibrium has been attained and determine the rise in the capillary tube. Was the membrane used a perfectly semipermeable membrane? Why or why not?

(B) Colloidal solutions show little or no osmotic pressure. However, it should be recalled that the stability of a hydrophilic colloid is dependent on two factors, namely, degree of solvation and an electrical double layer. If stability is dependent in part on an electrical double layer, then it signifies that ions adsorbed on the surface of the micelle are in equilibrium with the ions in true solution. The effect of the latter ions should be excluded if one is to measure the osmotic pressure of the dispersed particles. This can be accomplished by measuring the osmotic pressure of the dispersed particles against an ultrafiltrate of the colloid rather than against distilled water. Accordingly, prepare an ultrafilter as shown in Figure 1.2. Prepare a 1% bovine serum albumin solution (or any other hydrophilic colloid), allow to stand for 1-2 hr, then place the solution in the low pressure ultrafilter as indicated below and collect the ultrafiltrate. Prepare two cells using the albumin solution as outlined in (a) and determine the osmotic pressure of the same hydrophilic colloid in its ultrafiltrate and in distilled water. Compare. To what extent do colloidal particles exert osmotic pressure? Where would one expect a hydrophilic colloid to have a minimum osmotic pressure?

(C) Prepare a 1% gelatin solution. Test its pH. Add dilute acid or base to bring the solution to pH 4.7, the isoelectric point of gelatin. To one volume of solution add an equal volume of $N/500$ sodium hydroxide; to a second volume of solution add an equal volume of $N/500$ hydrochloric acid; and to a third volume add an equal volume of distilled water. Mix well and record the pH. Fill each osmometer with one of the above solutions and immerse in distilled water. Explain the results.

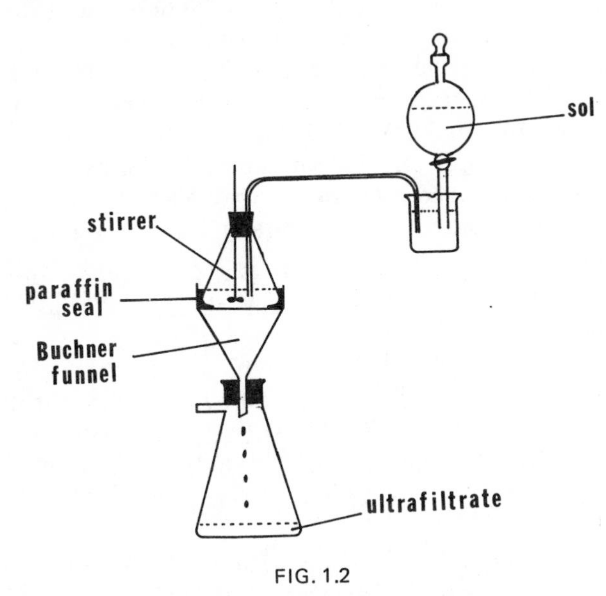

FIG. 1.2
ULTRAFILTER APPARATUS

Freezing Point Lowering

As pointed out above, whenever a solution is placed in contact with its solvent, diffusion occurs, i.e., there is a tendency for the solution to become more dilute until a uniform concentration is present in the entire system. If a frozen solvent, while at or very near its melting point, is brought into contact with a liquid solution at the same temperature, dilution occurs through melting of the solvent. Equilibrium can be restored by lowering the temperature of the system so as to increase the tendency of the solvent to solidify. Since freezing point depression is due to the attraction of solute molecules for the solvent, the amount of this lowering will be proportional to the number of solute molecules present (a colligative property).

The mathematical expression is:

$$\Delta t = K_f \times \frac{\text{wt solute}}{\text{mol wt (solute)}} \times \frac{1000}{\text{wt solvent}}$$

Where K_f = freezing point depression constant = lowering of freezing point by 1 mole of solute in 1000 g of solvent.

The above principle is used by dairy scientists to measure added water to milk. Milk has a relatively narrow freezing point range. Thus, if water is added to the milk, the milk becomes less concentrated and the freezing point rises from the normal range (-0.562° to -0.530°H) toward zero. Osmometers and/or cryoscopes are used for this determination.

Milk Cryoscopy

Materials:

(1) Cryoscope (Hortvet cryoscope or Advanced Instruments cryoscope)

(2) Standards:

(a) -0.422°H Standard - Weigh 7.0000 g NBS standard sucrose in 100 ml volumetric flask and dilute to volume with water, or weigh 100 g H_2O into 100 ml volume flask and add 0.6892 g of reagent grade NaCl.

(b) -0.621°H Standard - Weigh 10.0000 g NBS standard sucrose into 100 ml volumetric flask and dilute to volume with water, or weigh 100 g water in 100 ml volumetric flask and add 1.0206 g reagent grade NaCl.

Determination.—Apply the following technique in exactly the same manner for both standards and samples to obtain a valid freezing point value:

Advanced Milk Cryoscope.—Measure 2-3 ml sample and transfer to a clean, dry sample tube. Set measuring dial to the expected freezing point value. Place sample in cooling bath sample well, or in operating head, and lower operating head to position sample in cryoscope cooling bath.

Cool sample. Proper cooling is indicated by rapid and uniform movement of galvanometer needle from right to left over the scale of the meter (0.5-2.0 mm/sec).

Begin slow cooling at -11.5 to -2.0°H, depending on degree of supercooling desired.

Seed sample at -2.0° or -3.0°H, depending on make of cryoscope. Galvanometer spotlight shows the sample temperature rising to plateau.

Adjust galvanometer needle to coincide with the center line of the scale. Switch galvanometer to high sensitivity position. Keep centering the spotlight as it drifts to the right. Make the readout when the spotlight stops and remains temporarily on plateau before it begins moving to the left.

The measuring circuit is switched to low sensitivity. The sample tube is raised.

Determination of Added Water.—For instruments not calibrated for direct reading, the following formula may be used. First establish the Base Freezing Point (T) of authentic local samples. This value is used by the operator in reading or calculating percentage added water.

$$\% \text{ Added water} = \frac{100\ (T - T^1)}{T}$$

where
T = Base Freezing Point of authentic sample
T^1 = True Freezing Point of test samples

Interpretation: Freezing points colder than -0.530°H are not generally reported as violations.

CHEMICAL PROPERTIES OF SOLUTIONS

Acids and Bases

An acid (Bronsted) can ionize in water to form a proton and its conjugate base, while a base can react with a proton to form its conjugate acid.

$$HAc \Leftrightarrow H^+ + Ac^-$$
Acid Conjugate base

$$B + H^+ \Leftrightarrow BH^+$$
Base Conjugate acid

Thus, the acetate ion is the conjugate base of acetic acid and acetic acid is the conjugate acid of acetate; the couple CH_3COOH AND CH_3COO^- are called a conjugate acid-base pair:

$$CH_3COOH \Leftrightarrow H^+ + CH_3COO^-$$

When an acid is dissolved in water, some of the molecules or ions dissociate into H^+ ions and its conjugate base. The concentration of H^+ depends upon the degree of this dissociation. For example, acids such as HCl and HNO_3 and bases such as NaOH and KOH are almost completely dissociated while acetic acid and ammonia are incompletely dissociated ($K_i = 10^{-5}$). Clearly there will be intermediate examples, but the separation between strong and weak acids and bases may be placed at those acids and bases with dissociation constants (equilibrium constants) at 10^{-2} and 10^{-12} respectively.

The dissociation constants for weak acids and bases may be determined experimentally by preparing solutions of known concentration (c) and measuring the resulting pH. The equation for this approximation is:

$$K_a = \frac{[H^+]^2}{c}$$

or

$$[H^+] = \sqrt{K_a c}$$

Since $[H^+]$ can be determined with a pH meter and c is known, the dissociation constant (K_a) can be calculated.

One should be aware of the fact that the reaction of a solution can be expressed either in terms of the actual concentration of hydrogen ion in solution (intensity factor) or in terms of the total amount of acid or base present (quantity factor).

The intensity factor or actual concentration of hydrogen ions is a measure of the effective acidity of a solution. The concentration of hydrogen ion is most frequently expressed as pH.

The total amount of acid or base in a solution can be expressed in terms of normality or molarity. Titration is the procedure by which the total amount of acid or base is determined. By Le Chatelier's Principle, the equilibrium of the dissociation of a weak acid or weak base will be shifted by the removal of one of the components or by the addition of a common ion. For example, if H^+ is removed by titration with alkali then more HA will dissociate until finally no HA remains. Consequently, the amount of alkali required to titrate equal volumes of acid solutions of equal normality will be the same.

If to the above system the anion, A^-, were added in the form of a salt, the dissociation of the acid would be depressed and the $[H^+]$ would decrease. Thus, the addition of a completely ionized salt of the acid is similar to titrating the acid with a base. This is referred to as "common ion effect" and the relationship, known as the Henderson-Hasselbalch equation, can be expressed as follows:

$$pH = pK_a + \log \frac{\text{salt}}{\text{acid}}$$

A corresponding expression can be derived for a base from the dissociation constant of the salt of the base:

$$K_a = \frac{[B]\ [H^+]}{[HB^+]}$$

where B is the concentration of the base and HB^+ is the concentration of the protonated base. In terms of pH the relationship may be expressed:

$$pH = pK_a + \log \frac{[B]}{[HB^+]}$$

A buffer solution is one which resists changes in pH when small amounts of acid or alkali are added. The conjugate acid-base pairs of weak acids or weak bases act, by virtue of the mass law, as buffers, and as a consequence the pH of the buffer is related to the ratio of salt to acid. The following facts can be gleaned from the buffer equation:

$$pH = pK_a + \log \frac{[\text{salt}]}{[\text{acid}]}$$

(1) If the $[H^+]$ or pH is known, and K_a is known, the ratio in which one must mix the weak acid and its salts can be calculated.

(2) Vice versa, if the ratio of $\frac{[\text{salt}]}{[\text{acid}]}$ and K_a is known, then pH and $[H^+]$ can be calculated. (Usually $[H^+]$ is found to be larger experimentally than the calculations from the equation would suggest.)

(3) A buffer solution, having equal buffering capacity toward an acid and a base, will have a ratio of salt to acid of 1:1 and its $[H^+] = K_a$.

(4) If the ratio of salt to acid does not change, then the pH of the buffer solution should not change when diluted. If the ratio of salt to acid is unity and $M/10$ solutions of salt and acid are mixed in equal numbers, the pH of the solution will be approximately the same if the same ratio of $M/100$ solutions of acid is used. The former buffer will, however, have greater buffering capacity since it contains ten times the concentration of salt and acid as the latter. Thus, the former buffer solution has a higher buffer capacity which is defined as the mole equivalents of H^+ or OH^- that are required to change 1 liter of 1 M buffer by 1.0 pH unit.

Electrometric Titration

Materials:

(1) pH meter

(2) 0.3 M HCl; 0.3 M NaOH; 0.3 M HAc; 0.3 M NH_4OH

(3) Solid NH_4Ac

Procedure:

(1) Add 20 ml 0.3 M HCl to 250 ml beaker. Then, titrate potentiometrically with 0.3 M NaOH to the stoichiometrical end point.

(2) Note initial pH of the solution, the pH after adding increments of titrating solution (at least 29%, 50%, and 91% neutralized so that salt/acid is 1/10, 1/1 and 10/1) and the final pH of the solution.

(3) Plot a curve, with pH as the ordinate, and ml of acid or base added as the abscissa.

(4) Repeat for acetic acid versus sodium hydroxide; hydrochloric acid versus ammonium hydroxide.

Study Helps:

(1) Explain the differences in the graphs.

(2) In the titration, when one adds a weight of base equivalent to that of the acid present, is the end point of the titration at pH = 7? Why?

(3) What are some of the bits of information that can be derived from the curves?

Oxidation-Reduction Reactions

The term "oxidation" describes reactions involving a loss of electrons while "reduction" describes reactions involving a gain of electrons. An oxidation-reduction reaction always occurs simultaneously and is equivalent. The overall reaction can be expressed by the reaction:

$$\underset{\text{(Reductant)}}{\text{Electron donor}} \Leftrightarrow ne^- + \underset{\text{(Oxidant)}}{\text{Electron acceptor}}$$

where n is the number of electrons dissociated per molecule reductant.

The tendency of a substance to gain or lose electrons is determined by the nature of the substance and, as a consequence, it is necessary to have some standard electrode for comparison. By convention, the standard is the hydrogen electrode which has been assigned an oxidation-reduction potential E_o of 0.00 volts at pH 0 and 1 atm pressure. In quantitative terms, the following relationship exist:

$$E_{\text{obs}} = E_{\text{o}} + \frac{RT}{nF} \ln \frac{\text{oxidized form}}{\text{reduced form}}$$

where R is the gas constant (1.987 cal/°K/mole); T is degrees Kelvin; n is the number of electrons per gram equivalent transferred in the reaction; F is the Faraday (23,000 cal/volt·gram equivalent; E_{obs} is the observed potential difference in volts; and E_o is the standard oxidation-reduction potential of the electrode. The potential of the redox system varies with the ratio of oxidant to reductant and not with absolute concentrations.

If the reference electrode is a hydrogen electrode and the number of electrons transferred is unity at 30° C, then the above equation can be written as follows:

$$E_{\text{obs}} = E_o + 0.06 \log \frac{\text{oxidized form}}{\text{reduced form}}$$

The potential of H^+/H_2 redox system varies with the ratio of the components just as other redox systems. Consequently, from the above equation one can indicate the following relationship:

$$E_{\text{obs}} = E_o + \frac{0.06}{1} \log \frac{H^+}{\sqrt{P_{H_2}}}$$

Recall that in the standard state of H^+, which is one molar, the E_o is set at zero volts. Then

$$E_{\text{obs}} = 0.06 \log \frac{1}{\sqrt{P_{H_2}}}$$

Since H_2 is 1 and $-\log [H^+] = pH$, the above equation simplifies to:

$$E_{obs} = -0.06pH$$

In biological systems, it is convenient to compare data at pH 7.0 and 30°C. A secondary standard was adopted by which standard reduction potentials (E'_o) are expressed at pH 7.0,

$$\begin{aligned} E'_o &= -0.06 \text{ pH volt} \\ &= -0.06 \times 7.0 \text{ volt} \\ &= -0.42 \text{ volt} \end{aligned}$$

or

$$E'_o \text{ (at pH 7)} = -0.42 + E_o$$

It should be noted that any table of redox potentials for biochemical reactions which have been adjusted to condition of pH 7.0 will have the potentials reduced by -0.42 volt.

Systems having a more negative redox (E_o or E'_o) have a greater tendency to donate electrons. The more positive the redox potential the greater the tendency to gain electrons (acceptor).

Potentiometric Titration of Iron

Materials:

(1) A pH meter equipped with platinum/silver chloride electrode. N/10 Potassium Dichromate Solution— Dry a sample of $K_2Cr_2O_7$ for 2 hr at 160-180°, and cool in a desiccator. Weigh accurately 4.903 g of the dry sample and transfer to a 1 liter volumetric flask. Dissolve in water and dilute to mark.

(2) N/10 Ferrous Sulfate Solution. Weigh 28 g of crystallized ferrous sulfate, $FeSO_4 \cdot 7H_2O$, and transfer to a liter volumetric flask. Dissolve in water, add 5-10 ml of concentrated hydrochloric acid and dilute to mark. (Ferrous ion loses one electron when it is oxidized to the ferric state, the equivalent weight is its molecular weight.)

Procedure:

(1) Accurately measure 30 ml of the ferrous solution into a 400 ml beaker, add 100 ml of water and 20 ml of dilute hydrochloric acid solution (10 ml of concentrated HCl : 10 ml of water).

(2) Insert a platinum silver chloride electrode into the solution of Fe^{++} ion (or pH electrode and a reference calomel electrode). The voltage between the reference cell and the solution to be titrated is read on a potentiometer.

(3) Aliquots of $N/10$ dichromate are added and the potential difference measured.

(4) The titration is equivalent to the titration of weak acid with base except in the latter case the potentiometer is read in terms of pH rather than millivolts. The redox equation is also very similar to the Henderson-Hasselbalch equation and as a consequence the redox curves are similar in shape to the pH curves. (See previous page.)

QUESTIONS

(1) When is $\frac{RT}{nF}$ ln equal to 0.06 log?

(2) Is it necessary to know the pH of an oxidation-reduction system in order to report the oxidation-reduction potential?

(3) How does ln differ from $\log_{10}$?

(4) What is meant by the term "poise" of an oxidation reduction system?

(5) On what does the oxidation-reduction potential of a system depend?

(6) Will curves be parallel for all half-reactions with the same number of electrons in the equation and separated from each other by a voltage?

Colloids

VISCOSITY

Viscosity measurements are an estimate of the internal friction which viscous matter offers to deforming forces. Hydrophobic sols deviate but little from that of the dispersion medium while hydrophilic sols have viscosities which differ markedly from that of the dispersion medium. This property is of great importance with such food products as jams, jellies, gelatins, creams, catsup, oils, doughs, starches, syrups and mayonnaise. Viscosity is a quality attribute for frequently a product is accepted or rejected on the basis of the impression gained from the appearance of the product. Viscosity is also a useful measurement of the degree of hydrolysis of starch, pectin, and proteins. Viscosity can also serve as an index to the amount of an additive (e.g., gums) to be added to a food product, and also to heat treatment (protein denaturation) in a process which contains hydrophilic colloids. Viscosity can also be used in determining the moisture content of products (e.g., honey).

The absolute unit of viscosity is the poise. A poise, in cm-g-sec units, is the force which, when exerted on a unit area between two parallel planes 1 sq cm in area and 1 cm apart, produces a difference in velocity of streaming between the two planes of 1 cm per sec. The centipoise is 0.01 of a poise.

There are several types of viscosimeters available. Of these the most common are the Ostwald, Stormer, MacMichael and the Gardner instrument. The Ostwald viscosimeter, which is probably the simplest instrument, measures viscosity by flow through a capillary tube. Simplified versions of the Ostwald viscosimeter (e.g., the Jelmeter) are used by jelly manufacturers. The Stormer and MacMichael viscosimeters record the viscosity by shearing two solution layers which adhere to two surfaces 1 cm apart. The Stormer viscosimeter relates the viscosity to the time needed for a definite number of revolutions of a cylinder when immersed in a sample maintained at a definite temperature and actuated by a definite weight. Plots of time against weight used are made. The MacMichael viscosimeter records the angular torque of standard wires. The torque is a measure of the force needed to cause the two surfaces to move past each other at a velocity of 1 cm/sec. The Gardner instrument measures the time required for a weight to fall between two reference points in a tube of the material being tested.

Materials

(1) Viscosimeter
(2) Sugar (60% sucrose by weight)
(3) Gelatin (2.0% solutions). Place 4 g of gelatin in cold distilled water. Change the water frequently. To a weighed beaker add 150 ml distilled water and heat to boiling. Remove heat source, then add the washed gelatin. Disperse the gelatin with stirring; when the gelatin has been dispersed, cool and weigh the beaker plus contents, then add enough water to make 200 g of sol. Allow the sol to age in a refrigerator for 24 hr.

Procedure

(1) Viscosity vs Concentration.—Make a series of dilutions of the standard 60% sugar solution (10%, 20%, 40%). Temper at 20° C for 30 min. Determine the time of outflow of the prepared solution with an Ostwald pipet. Plot viscosity or time of outflow against concentration. The viscosity is a relative value because the viscosimeter must be calibrated against water, which has a value of 1.005 centipoise at 20° C. Prepare gelatin sols of 2, 1, 0.5, 0.25%. Fill the viscosimeter with the proper volume of the sol, then determine time of outflow. Work at constant temperature. Plot time of outflow as ordinate and temperatures as abscissa.

(2) Viscosity vs Temperature.—Prepare a 0.5% gelatin sol. Determine the viscosity of the gelatin solution at 0°C, 10°C, 20°C, 30°C and 40°C. Allow the filled viscosimeter to remain at the chosen temperature for 30 min prior to making the test. Plot time of outflow as ordinate and temperature as abscissa.

Repeat using water. Plot results on the same graph.

(3) Viscosity vs Electrolytes.—Prepare 600 ml of 2% gelatin solution as described above. Since the effect of salts on the viscosity of isoelectric gelatin is minimal it is preferable to study the effects of electrolytes on viscosity of gelatin solutions at pH's other than the IpH. Weigh six 2 g samples of gelatin and disperse in 90 ml of water. Adjust the six samples to pH 2, 4, 6, 8, 10, and 12, respectively, using either dilute HCl or NaOH (the final volume of the adjusted solution should not

exceed 100 ml). Adjust the final volume of each solution to 100 ml. Plot time of flow against pH.

Determine time of flow of 2% gelatin solution in various electrolyte solutions. Use *N* solutions of salts and add at rate of 1 ml/99 ml of gelatin solution. Note: Sulfates increase the gelatin viscosity greatly; phosphates, acetates and citrates increase the gelatin viscosity moderately; iodides decrease viscosity. Chlorides and nitrates may increase or decrease the viscosity of gelatin depending on their concentration.

(4) Viscosity of Emulsions.—Prepare emulsions containing 5, 10, 20, 40% of Nujol, or other oil, in 1% sodium oleate, lecithin or other commercial emulsifier solution. Measure time of flow against the concentration of oil in the emulsion.

The emulsion can be made with a laboratory homogenizer, or place 40 ml of the oil and 10 ml of a 1% water solution of sodium oleate, lecithin or other emulsifier in the emulsifier flask. Give the flask several vigorous shakes, then set aside for 30 sec. At the end of this rest period time, give it several more shakes. Continue this shaking after rest periods of 30 sec until the emulsion no longer separates. The time will be about 2.5-5 min. (If time permits shake another flask continuously for the length of time needed for intermittent shaking.) Will the emulsion continue to separate? What if only a 15 sec rest period is allowed: Will the elapsed time for the formation of the emulsion be greater than the 30 sec rest period?

QUESTIONS

(1) What are some of the factors which affect viscosity?

(2) What is the relationship between the IpH of a protein and viscosity?

(3) What is the difference between a viscous liquid and a plastic solid?

(4) Why does viscosity change with age?

(5) How might one distinguish between a lyophobic and a lyophilic sol?

Carbohydrates

Carbohydrates are essentially energy foods. They are divided into three main groups: monosaccharides, oligosaccharides and polysaccharides. Monosaccharides are simple sugars which cannot be hydrolyzed into simple compounds. They are the units which make up oligosaccharides and polysaccharides. Oligosaccharides yield two to six simple sugars on hydrolysis, while polysaccharides yield a large number of sugars on hydrolysis. In general, the more complex carbohydrates are hydrolyzed by enzymes to simple sugars, absorbed and oxidized in the body cells to yield energy. Excess body sugar (glucose) can be converted to glycogen and stored, or converted to fat and stored by the body. In either event, these reserves are later utilized by the body as a source of energy. There are, however, some polysaccharides such as cellulose, agar-agar, etc., which cannot be hydrolyzed by the enzymes of the digestive tract and are thus indigestible. The latter carbohydrates give bulk to the intestinal contents which is necessary for proper peristaltic action of the intestines.

QUALITATIVE ANALYSIS

It is sometimes necessary to determine the particular carbohydrates present in a food material prior to making a quantitative analysis. The kind of food will, in general, give some indication of the various carbohydrates to anticipate. For example, one would expect to find fructose and sucrose in uncombined forms in fruit juices and honey; milk products would contain lactose and, if sweetened, also sucrose; sweet potatoes would contain starch, sucrose and glucose. Specific carbohydrates may be detected by qualitative tests which depend mainly on differences in chemical structure. However, when employing these specific tests, it is frequently necessary to separate the particular carbohydrate from other carbohydrates in order to prevent interference with the test or arriving at erroneous conclusions due to the presence of other substances of a similar nature. This is particularly true for those tests involving the reducing action of the carbonyl group of sugars. In contrast, the starch-iodine sorption test is not appreciably affected by other substances and as a consequence it is a useful test for starch in the presence of large amounts of other substances.

GENERAL TESTS FOR CARBOHYDRATES

Reagents

Molisch's Reagent.—Dissolve 10.0 g α-naphthol in 100 ml of 95% alcohol.

Benedict's Reagent.—Dissolve 173.0 g sodium citrate and 100.0 g sodium carbonate in about 800 ml of water. Pour into a 1000 ml volumetric flask. Dissolve 17.3 g copper sulfate in approximately 100 ml of water. Add the copper sulfate solution to the flask with constant stirring. Then make to volume. Reagent does not deteriorate on long standing.

Barfoed's Reagent.—Dissolve 13.3 g of neutral, crystallized copper acetate in 200 ml of water. Filter if necessary, and add 1.8 ml of glacial acetic acid.

Seliwanoff's Reagent.—Mix 30 ml of water, in a 100 ml volumetric flask, with 60 ml of concentrated hydrochloric acid, add 0.5 g of resorcinol, and dilute to volume.

Iodine Solution.—Prepare a 2% solution of potassium iodide and add sufficient iodine to color it a deep yellow.

Orcinol Solution.—Dissolve 1.5 g of orcinol in 500 ml of hydrochloric acid and add 20 drops of a 10% ferric chloride solution. Prepare 0.5% aqueous solutions of the following carbohydrates: xylose, glucose, fructose, maltose, lactose, mannose, dextrin and starch.

Procedure

Molisch's Test.—The Molisch's test is based on the hydrolyzing and dehydrating action of concentrated sulfuric acid on carbohydrates. In the test the acid hydrolyzes any glycosidic bonds, if present, and dehydrates the monosaccharide to its corresponding furfural derivative. These furfurals then condense with α-naphthol to give a colored product.

Place 5 ml of each carbohydrate solution to be tested in a test tube, add 1-2 drops of the Molisch reagent, and mix. Incline the test tube and slowly run 5 ml of concentrated sulfuric acid down the side of the tube so that it forms a layer at the bottom of the tube without mixing. The formation of a reddish-violet ring at the interface of the two liquids indicates the presence of a carbohydrate.

Benedict's Test.—All monosaccharides as well as most of the disaccharide sugars possess the capacity to reduce: (1) alkaline solutions of copper, silver, mercury and bismuth, (2) acid solutions of molybdate and selenious acid and (3) organic compounds such as picric acid and methylene blue. In addition, they react with phenylhydrazine to form hydrazones and osazones. These sugars are referred to as reducing sugars because they have in their molecular structure a free (or potentially free) carbonyl group (aldehyde or ketone).

Place 1 ml of each carbohydrate to be tested in a test tube and add 5 ml of Benedict's reagent. Mix. Boil in a water bath for 2 min. A yellow or red precipitate of cuprous oxide indicates the presence of a reducing sugar.

Barfoed's Test.—The reagent used in this test is not reduced appreciably by disaccharides (lactose and maltose) but is reduced by monosaccharides. Thus, the test is useful in distinguishing monosaccharides in the presence of disaccharides.

Mix 5 ml of the reagent with 1 ml of the carbohydrate to be tested and place in a boiling water bath for 3-4 min. Examine for a red precipitate of cuprous oxide. (Dissaccharides may cause reduction if too much sugar or acid is present, or if the heating is too prolonged.)

Seliwanoff's Test.—This test was designed to differentiate between a keto sugar, more especially fructose, and an aldo sugar such as glucose or lactose. Sucrose, which contains fructose, also responds to the test.

To 5 ml of the carbohydrate solution add 5 ml of the resorcinol reagent and mix. Boil in a water bath for 20 min, cool quickly and examine after 2 min. A red color or red precipitate is a positive test for keto sugars.

Iodine Test.—This test for polysaccharides depends upon their reaction with iodine to form a blue-black starch-iodine complex. Starch, most dextrins, amylodextrins and glycogen will give a positive test.

Place 2 ml of each carbohydrate solution in a test tube and add 3 drops of the iodine solution. Unbranched macromolecules (amylose moiety) give a blue color and branched macromolecules (amylopectin) give a reddish-black color.

Bial's Test.—This is a characteristic test for pentoses or carbohydrates capable of yielding pentoses.

Heat 5 ml of the reagent to boiling, remove from the flame, and add a few drops (less than 1 ml) of the carbohydrate solution to be tested. The presence of a pentose is evidenced by a vivid green color which develops almost immediately.

Detection of Sugars by Paper Chromatography

Paper chromatography is a useful method for qualitatively identifying sugars, especially in those instances where the amounts of sugar present in the solution are relatively small. There are a variety of solvents for developing the chromatograms. Most sugars can be separated on filter paper by a phenol-water solvent. In general, the order of R_f values is: pentoses (highest), hexoses, disaccharides and trisaccharides. There are a variety of methods available for quantitative analysis of sugars by paper chromatography.

Materials

(1) Phenol-water solution: 80 parts phenol are mixed with 20 parts water (80:20).

(2) Whatman No. 4 paper.

(3) Sugar solutions (0.1 *M*): glucose, ribose, raffinose, fructose, xylose, sucrose, maltose.

(4) Spray indicators.

(a) Reducing sugars (Tollen's reagent). Equal parts of 0.1% silver nitrate and 5 *N* ammonium hydroxide are mixed prior to spraying.

(b) Nonreducing sugars (Anisidine hydrochloride). Three grams of *p*-anisidine hydrochloride are dissolved in 100 ml n-butanol.

Procedure

(1) Draw a pencil line 2 cm from and parallel to the edge of a 20 × 20 cm piece of Whatman No. 4 filter paper. Along this line, at intervals of 2 cm, mark a small dot.

(2) Using a 20 microliter micropipet (Dispo-Scientific Products Co.) spot quantities of sugar solution on the dot, taking care that the spot remains small and discrete. Allow to dry for 10 min.

(3) A developing jar (or 4 liter beaker) is prepared by placing 1/2 cm depth of developing solvent into the jar.

(4) The dried filter paper is formed into a cylinder, which is stapled top and bottom, taking care to see that the edges of the sheet are not touching. A paper clip is used to join the center portion of the cylinder. The cylinder is now ready for development.

(5) The paper cylinder is placed in the developing chamber and the developing chamber is then closed with a cover plate. (Saran wrap and a rubber band may be used to close the tank if a cover plate is not available.) The solvent is allowed to proceed up the paper (ascending chromatography) to within 2 cm of the top. This should take approximately 6-9 hr.

(6) Following removal of the chromatogram, the solvent boundary on the upper portion is rapidly

marked and the paper is allowed to dry under a hood. If the boundary cannot be adequately defined, allow the chromatogram to dry and view under ultraviolet light. The boundary will be perceptible.

(7) After drying, spray lightly with Tollens reagent, then warm for a few minutes. Warning! Do not overheat or the paper will turn black. This spray is suitable for identifying reducing sugars. Alternatively, the chromatogram may be sprayed with *p*-anisidine hydrochloride and warmed 5-10 min at 105° F. This spray will detect both reducing and nonreducing sugars.

QUANTITATIVE ANALYSIS

There are four general types of methods which may be utilized for the quantitative analysis of sugars in foodstuffs. They are: (1) reduction methods, (2) refractometric methods, (3) polarimetric methods, and (4) densimetric methods. Previous to the use of these quantitative methods, representative samples had to be obtained and the sugar-containing solution clarified because the solution contains, in addition to soluble sugars, substances which will interfere with the analysis. The interfering substances are soluble pigments, optically active substances (amino acids, etc.), phenolic constituents, lipids and protective colloids (protein). The interfering substance may be separated by decolorization, ion exchange resin treatment, or clarification with various clarifying agents (alumina cream, lead acetates, phosphotungtic acid, animal charcoal). (See Joslyn, *Methods in Food Analysis*, or AOAC 1975.)

ANALYSIS OF A SUGAR PRODUCT (MAPLE SYRUP)

Maple syrup is obtained by concentrating the sap from hard or rock maple trees. Maple syrup is required to contain no more than 35% water and must weigh at least 11 lb per gallon at 60° F. The principal sugar of maple syrup is sucrose (1-6%) but it contains organic acids, mineral matter, proteins and flavoring materials characteristic for maple syrup.

Materials

(1) Maple syrup refractometer
(2) Hydrometer
(3) Neutral lead acetate. Prepare a saturated solution of neutral lead acetate. (See AOAC, 1970, 31.021.)
(4) Hydrochloric acid solution. Prepare 100 ml of a hydrochloric acid solution with a sp gr 1.103 @ 20° C
(5) Polarimeter
(6) Asbestos (See AOAC, 31.038.), Amphibole variety
(7) Fehling's A & B solution
(8) Alcohol
(9) Ether

Procedure

Refractometric Method.—The refractive index of a sugar solution is a direct measure of its concentration. Solutions of different sugars of equal concentration have approximately the same refractive index. The speed and ease with which the refractive index of a sugar solution can be determined makes a convenient method for determining the sugar content of solutions, and indirectly the water content of sugar solutions. Consequently, the refractometer is widely used for quality inspection in the manufacture of syrup, jams, fruit juice and other food products. Determine the refractive index of the syrup at 20° C with either an Abbe or hand refractometer. Obtain the corresponding percentage of soluble solids (as sucrose) from a table in the AOAC relating Refractive Indices to Sucrose percent (AOAC 1970, 47.014).

Densimetric Method (Hydrometer Method).—The specific gravity of a sugar solution is a function of the sugar concentration (solute) at a definite temperature. The presence of other soluble materials will affect the specific gravity of the solution. However, in relatively pure sugar solutions the method will give a fair approximation of the amount of sugar present in the solution. The hydrometer is also widely used for the measurement of soluble solids. A special hydrometer, reading in percentage of sugar directly at 20° C (degrees Brix), has been developed for sugar work (AOAC 1970, 31.009). The solution (at 20° C) is placed in a tall cylinder and the Brix reading obtained after the hydrometer is spun and comes to rest.

Polarimetric Method.—This method depends upon the fact that sugars contain asymmetric carbon atoms and hence have the ability to rotate a plane of polarized light through an axis parallel to its direction of propagation. The angle through which this rotation occurs is directly proportional to the concentration of the sugar, the length of the tube through which the light passes and the specific rotating power of the sugar. The polarimeter may be used for the quantitative determination of sugars but in actuality the saccharimeter is more commonly employed. The essential differences between these two instruments are that the polarimeter employs monochromatic light and reads in angular degrees, whereas the saccharimeter employs white light and a quartz wedge compensation so that the percentage of sugar may be read directly. The saccharimeter reads sugar concentration directly, provided a single sugar is present, and provided also that a normal weight of sugar is used for the reading. A normal weight of sugar is defined as that weight which, when made to volume of 100 ml and viewed in a 200 mm tube at 20° C[1], will give a reading of 100.

The principal sugar of maple syrup is sucrose. However, a small amount of invert sugar is normally present due to the hydrolysis of a portion of the sucrose. The Clerget-Herzfeld saccharimetric method may be used for the determination of sucrose in maple syrup. This method involves two readings. The first reading, based on a normal weight of sample, is due to the sucrose and invert sugar in the sample. The second reading of a normal weight, after inversion of the sample, is due to the invert sugar derived from sucrose plus the invert sugar originally present. The change in rotation following hydrolysis is a function of the amount of sucrose present in the sample. The change of rotation of a normal sugar following hydrolysis can be found in the literature. Use of this value in the proper equation then allows one to calculate the amount of sucrose in the sample.

The AOAC procedure (AOAC 1970, 31.026) using the saccharimeter is as follows: Rapidly weigh to the nearest 0.005 g, 52.00 g of the sample (twice the normal weight) in a dish and transfer with water to 200 ml volumetric flask. Add 2-5 ml of neutral lead acetate solution, dilute to volume and mix. Filter and discard the first 25 ml of filtrate. Remove the excess lead from the filtrate by adding anhydrous sodium carbonate, a little at a time, avoiding an excess; mix well and filter again, discarding the first 25 ml of filtrate.

Direct Reading.—Pipet a 50 ml aliquot of the lead-free filtrate into a 100 ml volumetric flask, add 2.315 g NaCl, dilute to volume with water at 20° C and mix well. Fill a 200 mm saccharimeter tube with this solution and obtain a reading in the saccharimeter at 20° C. Twice the normal weight in a 200 ml volume is equivalent to a normal weight in a 100 ml volume. A 50 ml aliquot diluted to 100 ml then represents a sample of one half the normal weight. Thus the saccharimeter reading obtained above when multiplied by two will give the reading based on a normal weight. (Save the remainder of the solution if one wishes to make a gravimetric determination of the percentage of invert sugar.)

Invert Reading (Inversion At Room Temperature).—Pipet a 50 ml aliquot of the filtrate from the original 200 ml of clarified and deleaded solution into a 100 ml volumetric flask, add 20 ml H_2O and 10 ml HCl (sp gr 1.103) and let stand for 24 hr at a temperature = 20° C. If the temperature is above 28° C, a 10 hr digestion is sufficient. Dilute to volume and determine the saccharimeter reading at 20° C. Again, the 50 ml aliquot represents a sample of one half the normal weight, and, as a consequence, the reading obtained must be multiplied by 2. This will give the saccharimeter reading based on a normal weight. The sucrose present may be calculated from the formula:

$$S = \frac{100(P-I)}{132.56 - 0.0794(13-m) - 0.53(t-20)}$$

where

S = % sucrose

P = direct reading on a normal solution

I = Invert reading on a normal solution

m = grams of total solids from the original sample in 100 ml of the inverted solution.

t = temperature at which readings were made

Gravimetric Determination of Invert Sugar (Munson-Walker Determination).—The analysis depends upon the reducing action of sugars with free carbonyl groups. Since it is an empirical procedure it must be followed exactly to obtain comparable results.

Shake amphibole asbestos with distilled water until a fine pulp is obtained. Prepare a Gooch crucible by forming a mat of asbestos about 1/4 in. thick. Wash with 10 ml alcohol and then with 10 ml ether. Dry 30 min at 100° C, cool in a desiccator and weigh. Transfer 25 ml each of Fehling's solution A and B to a 400 ml beaker. Add 50 ml water, cover with watch glass and heat on asbestos gauze over a Bunsen burner. Regulate flame so that boiling begins in 4 min and continue boiling for exactly 2 min. Filter at once through the prepared Gooch crucible. Wash the precipitate (if any) thoroughly with hot water (about 60° C), then with 10 ml alcohol and finally with 10 ml ether. Dry in an oven at 100° C for 30 min, cool and weigh. Record weight increase (blank value).

Pipet 25.0 ml of the solution used for obtaining the direct reading on the saccharimeter into a 400 ml beaker. Add 25 ml of water and 25 ml each of the two Fehling's solution. Cover beaker with a watch glass. Heat beaker on asbestos gauze over a Bunsen burner. Regulate flame so that boiling begins in 4 min and continue boiling *exactly 2 min.* Filter hot solution at once through a Gooch crucible equipped with an asbestos mat, using suction. Wash precipitate of Cu_2O thoroughly with water at about 60° C, then wash with 10 ml alcohol, and then with 10 ml ether. Dry in oven at 100° C for 30 min, cool and weigh. Subtract

[1] Temperature must be considered because sugars tend to show a decrease in specific rotation with an increase in temperature. In contrast, fructose, galactose and invert sugar show an increase in specific rotation with increase in temperature; sucrose and glucose show little or no change in specific rotation with increase in temperature.

weight of blank, then obtain the corresponding amount of invert sugar from the table headed "Invert Sugar and Sucrose, 2 g of Total Sugar" (AOAC 1970, pg. 938). This is necessary to correct for the reducing action of the sucrose present in the sample. Calculate the precentage of invert sugar present in the original syrup on the basis of the aliquot used.

HYDROLYSIS OF STARCH

Many different kinds of carbohydrates occur in foods but not all are of equal importance as suppliers of energy. Of the polysaccharides, starch is the principle one that man can use efficiently. All carbohydrates must be broken down to monosaccharides before they can be absorbed and utilized by the body.

Analysis of starch in various foods generally requires either acid or enzymatic hydrolysis or a combination of both. In the following procedure a combination of both techniques is utilized.

Reagents and Equipment

(1) α-Amylase—Dissolve 10 mg of commercial enzyme (Sigma Chemical Co.) in 20 ml of pH 6.9 phosphate buffer. The enzyme should have an activity such that one gram of the enzyme will digest 50 g of starch in 30 min under standard conditions.

(2) Amyloglucosidase—Dissolve 100 mg of commercial enzyme in 2.0 ml of pH 4.3 acetate buffer. One unit of activity will yield 1.0 mg of glucose from starch in 3 min at 55°C and pH 4.5. Commercial enzyme prep contains 1200-3000 units per gram of enzyme.

(3) pH 6.9 Phosphate Buffer—Dissolve 170 mg Na_2HPO_4 (anhydrous), 140 mg $NaH_2PO_4 . H_2O$, and 30 mg NaCl in 100 ml H_2O. The pH of the resulting solution then needs to be checked and adjusted to pH 6.9 by adding NaOH or HCl. This pH is very critical and must be exact! Unless the exact pH specifications are adhered to, the enzymes will lose their ability to breakdown starch to glucose.

(4) pH 4.3 Acetate Buffer—To prepare 500 ml, use 29.4 ml of glacial acetic acid (17 *N*) in 470.6 ml of H_2O. Add sodium hydroxide to this solution until an *exact* pH of 4.3 has been reached. Be sure to standardize the pH meter each time you check the pH of different solutions. Use standard pH solutions to standardize the meter. For example, to check a pH 4.3 solution, standardize with a commercial pH 4.0 buffer.

(5) 2 *N* HCl—To prepare 200 ml, mix 33 ml concentrated HCl with 167 ml H_2O.

(6) 3 *N* NaOH—To prepare 200 ml, mix 24 grams NaOH with 200 ml H_2O.

(7) 3,5—Dinitrosalicylate Reagent—To prepare 500 ml, dissolve with warming, 5 grams 3,5-dinitrosalicylic acid in 100 ml 2 *N* NaOH (8 g NaOH in 100 ml H_2O). Add 150 g sodium potassium tartrate to 25 ml H_2O, and warm to dissolve. Mix the two solutions, and make up to 500 ml with H_2O.

(8) Glocose standard.

(9) pH meter, hot plate, mechanical stirrer.

Procedure

Preparation of the Corn Samples and Acid Pretreatment.—Grind the corn in a Wiley mill or mortar. Weigh out exactly 0.80 g of the corn and place in a 250 ml beaker. Add 25 ml of 2 *N* HCl and approximately 25 ml of H_2O to permit more dissolution. With a mechanical stirrer agitate the solution for 20 min. Place on a preheated hot plate (around 200°-300°C) and bring to boil. After 20 min remove the corn solution and cool to room temperature. Using a pH meter slowly add 3 *N* NaOH until a pH of exactly 6.9 is reached. If the pH is overshot, 2 *N* HCl may then be added to adjust it back to pH 6.9. The volume of this solution should now be approximately 100 ml. Place solution in 100 ml volumetric flask and add H_2O to volume. From this solution (which contains some cellular matter which has little effect upon the final results) a 0.4 ml aliquot is withdrawn for each subsequent analysis.

Enzyme Hydrolysis.—Add exactly 0.75 ml of the α-amylase reagent to a 0.4 ml aliquot of the acid hydrolysate. (Avoid contact of the enzyme reagent with the upper regions of the test tube.) Allow mixture to react for at least 3 min (no more is needed but 3 min is necessary). Then add exactly 0.10 ml of amyloglucosidase reagent. The resulting pH should then be exactly 4.5. At this pH the maximum starch hydrolysis is obtained from amyloglucosidase. The samples are then placed in a water bath at 55°C (±1°C) for 10 min.

Color Development.—Add 1.0 ml of 3,5-dinitrosalicylate reagent. (Again avoid contact with the sides of the tube.) Prepare a blank by adding 1.0 ml of the 3,5-dinitrosalicylate reagent to 1.25 ml of water. Place the test tubes into a boiling water bath for exactly 5 min. Remove the tubes and cool to room temperature. Dilute the contents with H_2O to 20 ml and mix. Using the blank determine the absorbence at 540 nanometers.

Concurrently with the above samples a standard curve should be prepared using glucose standards in range of 1 mg to 5 mg.

Standard Curve for Various Glucose Concentrations.—A standard curve of absorbence versus glucose concentrations is essential in determining the number of milligrams of starch present in any sample of corn. Absorbence is plotted as the ordinate and the concentration of glucose (in milligrams) is plotted as the abscissa. Absorbence is located on the Y-axis; then go horizontally across the graph paper until you intersect the line through the points. From this point, go straight down until the X-axis is intersected. This is the value in milligrams of the starch in the corn sample. To find the percent starch use the formula:

$$\% \text{ starch} = \frac{\text{wt starch in sample} \times 100}{\text{wt sample}}$$

QUESTIONS

(1) Why does "inversion" occur when sucrose is hydrolyzed?

(2) Show reactions of sugars that form furfural and hydroxymethyl furfural.

(3) What is the function of α-amylase and amyloglucosidase in the determination of starch in corn?

(4) Do monosaccharides contain a free aldehyde group? Explain.

ANALYSIS OF SUGARS BY GAS CHROMATOGRAPHY: TMS-SUGAR-OXIMES[1]

The use of TMS-Sugar-Oxime derivatives offers specific advantages for the analysis of sugars in food products, syrups, etc., which contain mixtures of fructose, glucose, sucrose, maltose and lactose. Trimethylsilyl ether derivatives of sugars are sufficiently volatile to allow gas chromatographic analysis. In the method, sugars are treated with hydroxylamine hydrochloride and the resulting oximes are converted to the trimethyl (TMS) ethers by the addition of silylating agent. The procedure offers the following advantages:

(1) Fast, single vial reaction of 1-1/2 to 2 hr analysis time. Oximes are silylated directly. There is no isolation of oximes prior to silylation (see Procedure).

(2) Single peaks simplify chromatograms and calculations. Multiple peaks, due to the tautomeric forms of reducing sugars, are largely eliminated, and single peaks are obtained with fructose, glucose and maltose.

(3) Good separations, especially useful in separating fructose and glucose.

(4) Direct analysis of aqueous solutions. Procedure is tolerant of up to 20 mg water present in each aliquot of sample being derivatized. Syrups containing 20-30% or more water need not be concentrated prior to analysis as long as no more than 20 mg of water are present in sample being derivatized. Samples containing 70-75% water should be concentrated.

(5) Quantitative and reproducible. Multiple chromatograms of known solutions should have an accuracy of ±3%.

Reagents and Supplies

1. Sugars or syrup
2. STOX oxime reagent with internal standard. (Available from Pierce Chemical Co., Rockford, Ill. Instructions are furnished with the reagent.)
3. Hexamethyldisilazane
4. Trimethylsilylimidazole
5. Chromatography columns: 6′ × 1/8″ O.D., S.S. packed with 2% or 3% OV-17 on Chromosorb W(HP) 80/100 mesh.

Procedure

Derivative Formation and GC Analysis

(1) Accurately weigh 10-15 mg (dry basis) of sugar or syrup mixture into a 3.5 ml screw cap septum vial.

(2) Add 1.0 ml of "STOX" oxime-internal standard reagent. Heat for 30 min at 70-75°C. Proceed to 3a or 3b.

(3-a) Cool to room temperature and add 1.0 ml hexamethyldisilazane, mix, and add 0.1 ml trifluoracetic acid. Cap vial and shake for 30 sec. Allow to stand at room temperature for 30 min. (White precipitate appears which will settle out.)

(3-b) Cool to room temperature and add 1.0 ml trimethylsilylimidazole. Cap and shake for 30 sec. Allow to stand at room temperature for 30 min (no precipitate).

(4) Analysis an aliquot on a 6′ × 1/8″ O.D., S.S. column packed with 2% (Fig. 3.1) or 3% (Fig. 3.2) OV-17 on Chromosorb™ W (HP) 80/100 mesh. Inject sample and program immediately from 150°-325°C at a rate of 10°C per min. When analyzing for fructose, glucose and sucrose, using 3% OV-17, a program from 140°-250°C with an initial 2 to 3 min hold before programming is sufficient to separate all sugars. Carrier gas-helium at 40 ml per min FID detector.

QUESTIONS

(1) Write equations for the reaction between sugar oximes and the silylating agent.

(2) What advantages does the TMS-Oxime method offer over other methods such as paper or thin-layer chromatography?

(3) Why are the sugars first converted to oximes before reaction with the silylating agent?

[1] Pierce Chemical Company, Rockford, Ill.

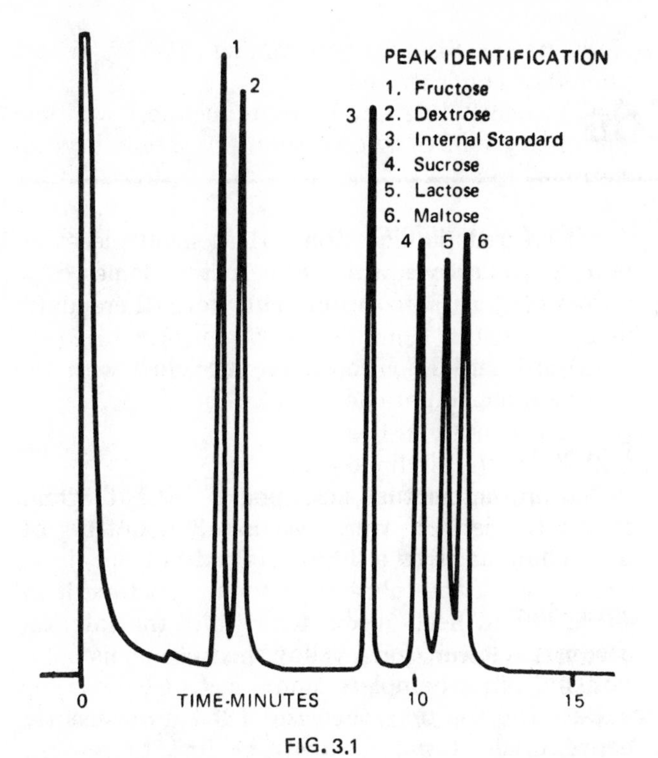

FIG. 3.1

GAS CHROMATOGRAPH OF SUGARS AND SUGAR STANDARD ON 2% OV-17

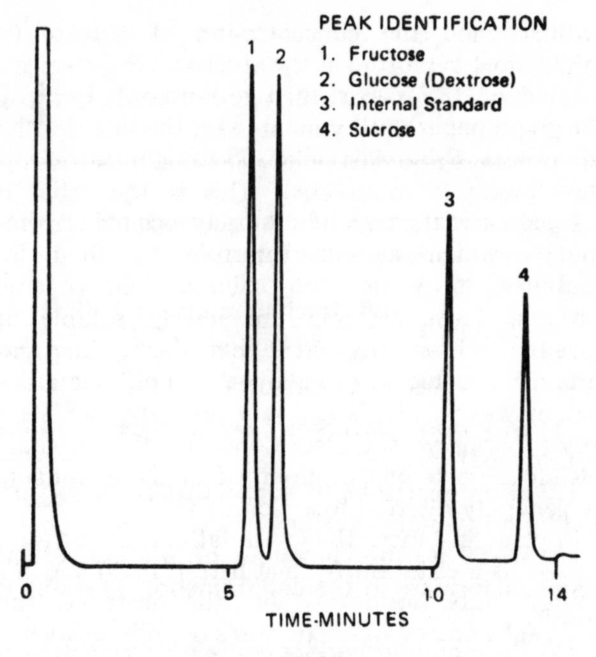

FIG. 3.2

GAS CHROMATOGRAPH OF SUGARS AND SUGAR STANDARD ON 3% OV-17

Lipids

Lipids are a group of naturally occurring compounds which are characterized by their insolubility in water and solubility in organic solvents. Fats and oils are readily soluble in solvents such as ethyl ether, petroleum ether and carbon tetrachloride, while other lipid materials such as waxes, sterols and fat-soluble pigments are only sparingly soluble. Consequently, organic solvent extracts of this group of food components are generally referred to as "crude fat."

In some instances the crude fat content of food is used as a criterion for maturity (olives) and for quality (nuts, soybeans). In other instances the crude fat content of foods poses definite problems to a food processor because degradation of lipids gives rise to off-flavors.

CHEMICAL AND PHYSICAL PROPERTIES OF FATS AND OILS

Fats and oils are identified by their chemical and physical constants. Of the chemical constants, the most widely used are the iodine and Reichert-Meissl numbers. Instrumental methods of analysis, such as gas-liquid chromatography, are more commonly used, but it was deemed advisable to introduce the older acceptable chemical methods. Of the physical constants, the most widely used constants are specific gravity, refractive index and melting points.

Chemical Constants

Reichert-Meissl Number.—This value may be defined as the number of milliliters of tenth normal alkali required to neutralize the volatile water-soluble acids obtained from 5 g of fat. A high number, therefore, indicates the presence of low-molecular weight fatty acids, particularly butyric acid and to a lesser extent caproic and caprylic acids. This determination has been used principally for the analysis of butter.

Materials

(1) 10-15 g of either butter fat, oleomargarine or hydrogenated vegetable fat

(2) Glycerol-soda solution. Add 20 ml of a clear saturated solution of sodium hydroxide to 180 ml of glycerol

(3) Dilute sulfuric acid solution. To 4 volumes of water add 1 volume of concentrated sulfuric acid

(4) Treated pumice stone. Heat small pieces of pumice stone to a white heat in the flames of a burner, plunge into water, and keep there until used

(5) Standard 0.1 *N* sodium hydroxide

(6) 1% phenolphthalein in alcohol

Procedure

Set up apparatus according to the diagram shown in Fig. 4.1. Weigh accurately about 5 g of the melted fat into a 300 ml round bottom flask, add 20 ml of the solution of sodium hydroxide in glycerol, and heat over a flame until the solution becomes a clear, pale yellow homogeneous solution, indicating complete saponification.

Cool the solution, then add 135 ml of recently boiled distilled water, slowly at first to prevent foaming. Add 10 ml of the dilute sulfuric acid solution, shake the mixture well and add a few pieces of the treated pumice stone.

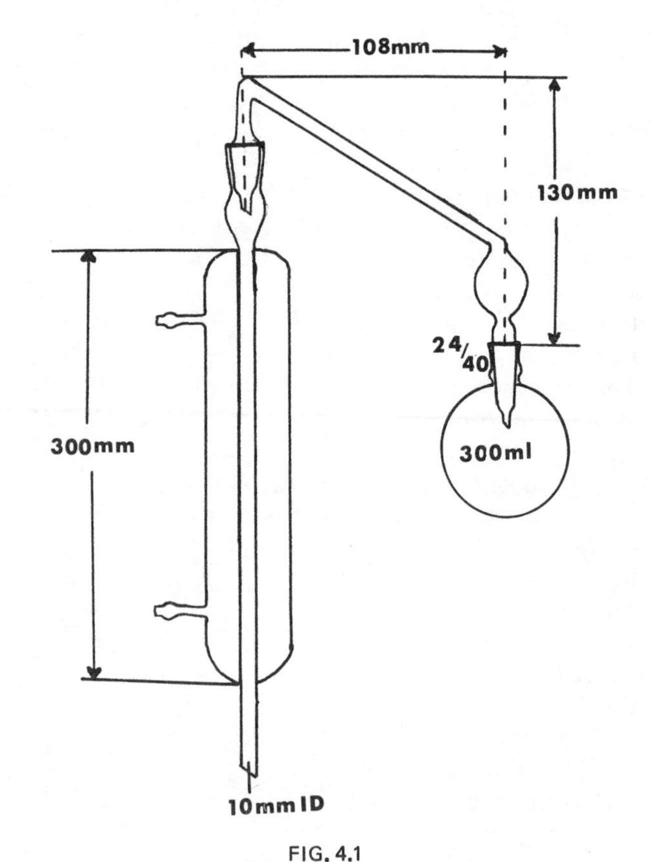

FIG. 4.1
APPARATUS FOR DISTILLING LOW-MOLECULAR WEIGHT FATTY ACIDS

Distill the volatile fatty acids in an apparatus having the approximate dimensions illustrated in Fig. 4.1. Rest the flask on a piece of asbestos board having a 5 cm hole in the center. Regulate the flame so as to obtain 110 ml of distillate in approximately 30 min. The distillate should not go above 20°C when it dips into the receiving flask. When the distillation is complete, remove the flame and allow to stand for a few minutes to collect the few drops of distillate which come over after the flame has been removed.

Shake the distillate gently, then immerse the flask in a water bath at 15°C for 15 min (to solidify any water-insoluble acids that have been carried over by the steam). Filter (why?) the entire distillate through a dry, 9 cm filter paper, and titrate a 100 ml aliquot with standardized 0.1 *N* sodium hydroxide, using phenolphthalein as an indicator. A blank determination, using all reagents and the same technique as was used on the samples, is run simultaneously with the samples.

The Reichert-Meissl number is the titration figure obtained, calculated to a 5 g fat sample, multiplied by 1.1 (remember that only 100 ml of distillate was titrated) and corrected for the blank titration. The determination of the Reichert-Meissl number of a fat is empirical because only a fraction of the volatile fatty acids are distilled over with steam and are collected in the 110 ml of distillate (80% butyric, 25% caprylic). Consequently, the directions must be followed as completely as possible.

Iodine Number of Fats

The iodine number is defined as the number of grams of iodine absorbed by 100 g of fat. The value obtained is a measure of the degree of unsaturated fatty acids in the fat. Butterfat and oleo oil have comparatively low iodine numbers (28-40). Practically all edible fat and vegetable fats have iodine numbers ranging from 65-130. Consequently, the iodine number of a glyceride is useful in determining adulteration of butterfat or oleo oil.

For this determination use a portion of the same lot of fat as was used in the above experiment. Do duplicate determinations on the triglyceride and the reagents. The amount of halogen reacting with the unsaturated fatty acids of the glyceride is calculated as iodine.

Materials

(1) Four 300 ml glass stoppered flasks

(2) Two 50 ml burettes

(3) Chloroform

(4) 10 g butterfat or oleomargarine fat

(5) Wijs iodine solution. Dissolve 13 g of resublimed iodine in 1 liter of glacial acetic acid using heat to promote a more rapid solution of the iodine. Cool the solution and titrate a 20 ml portion with 0.1 *N* sodium thiosulfate solution. Pour a portion of the prepared solution into a beaker and retain. Pass dry chlorine gas into the main portion of the solution until the original titration figure is doubled. Pour the small portion of the original iodine solution into the solution that was chlorinated. This should reduce the chlorine content of the entire solution to less than half of the iodine content

(6) Standardized 0.1 *N* sodium thiosulfate. Weigh out accurately 1.2-1.4 g of pure anhydrous sodium thiosulfate, transfer to a 250 ml volumetric flask, dissolve in distilled water and dilute to the mark

Place 100 ml boiled distilled water in a 500 ml Erlenmeyer flask, add 5 g of potassium iodide and mix with a gentle rotary motion until dissolved. Add to the latter solution, with mixing, 2.0 g of sodium carbonate, then add 7-8 ml of concentrated hydrochloric acid.

Remove a 25 ml aliquot of the potassium dichromate solution and slowly add it to the flask containing the potassium iodide solution. Mix the solution with a rotary motion, cover the Erlenmeyer flask with a watch glass or inverted beaker, and let stand in the dark for 5 min. Remove the flask from the dark and dilute solution to a volume of approximately 300 ml with distilled water. Titrate immediately with the thiosulfate until the yellow color of the iodine becomes faint, then add 3 ml of the starch indicator and continue the titration, drop by drop, until the blue-green color changes to light green.

Calculate the normality of the sodium thiosulfate solution from the above data and the relationship given in the following equations:

$$K_2Cr_2O_7 + 6\,KI + 14\,HCl \rightarrow 2\,CrCl_3 + 8\,KCl + 3\,I_2 + 7\,H_2O$$

$$I_2 + 2\,Na_2S_2O_3 \rightarrow Na_2S_4O_6 + 2\,NaI$$

(7) A potassium iodide solution containing 15 g of potassium iodide dissolved in 85 ml of water

(8) Starch indicator solution. Add 1 g of soluble starch to 10 ml water and stir, then add the well mixed suspension into 100 ml of boiling water and continue the heating for 2-3 min. Allow to cool and use as the indicator solution

Procedure

Weigh accurately, to the third decimal place, about 0.5 g of the fat into a glass-stoppered iodine flask. Add 15 ml chloroform as a solvent to each flask. From a burette, add 25 ml of the Wijs iodine solution, stopper the flask and place in the dark for 30 min. Shake occasionally. At the end of the 30 min period, add 20 ml of the 15% potassium iodide solution; shake thoroughly, then add 100 ml of freshly boiled and cooled distilled water, washing into the flask any free iodine that may be present on the stopper. Titrate immediately with standard 0.1 *N* sodium thiosulfate solution, with constant shaking until the solution has become light yellow in color. Add a few drops of the starch indicator solution and continue the titration, drop by drop, shaking vigorously after each addition, to the disappearance of the blue color.

Blank determinations are run with the samples. The same quantities of reagents are used as in the unknowns and the volumes are measured with the same volumetric apparatus. The average value for the blank determinations is a measure of the halogen concentration of the Wijs iodine solution.

Calculations:

Calculate the Iodine Number using the following formula:

$$\text{Iodine Number} = \frac{(\text{Blank titration} - \text{Sample titration}) \times \text{Normality of } Na_2S_2O_3 \times 12.69}{\text{Sample weight in grams}}$$

Physical Constants

The physical methods used in identifying glycerides are so well known it seems hardly necessary to make remarks on them, yet for completeness in this discussion some of the methods will be considered as experiments.

Refractive Index.—The refractive index may be defined as the ratio of the speed of light in air to its speed in the medium in question or briefly:

$$\text{Refractive Index} = \frac{\text{Speed of light in air}}{\text{Speed of light in a given medium}}$$

That speed of light varies with the medium through which it is passed is at once evident to anyone who has looked at the appearance of a glass rod in a beaker of water. The water being more dense than air slows the speed of light, with the result that light from that part of the glass rod in the water reaches the eye a fraction of time later than light from that part of the rod in the air. This characteristic presents a simple yet accurate way for the rapid determination of the purity of a fat. The refractive index can be obtained by means of some type of refractometer (e.g., Abbé refractometer). The reading is normally made at 40°C, or corrected to that temperature using the constant 0.000365 change in refractive index per degree C change in temperature.

Specific Gravity.—The specific gravity of a substance is the ratio of the weight of a given volume of a substance to the weight of a like volume of water at a specified temperature. The specific gravity of a glyceride can be determined very quickly by means of a hydrometer. Hydrometers for use with fats and oils should measure specific gravities below one, the value of water, since fats and oils are lighter than water. The reading is normally made at 15.5°C (60°F).

Materials

(1) One hydrometer reading from 1.000 to 0.850

(2) One cylinder (300 ml)

(3) 250 g each of olive oil, cottonseed, corn and soybean oil cooled to 15.5°C

Procedure

Nearly fill the cylinder with the oil cooled to a temperature of 15.5°C. Introduce the hydrometer and when the hydrometer comes to rest read the scale at the point where it is level with the surface of the oil. Record the reading.

Melting Point.—This value serves as an indicator of the kinds of fatty acids in a triglyceride. Unsaturated and low-molecular-weight saturated fatty acids make for low melting points; saturated high-molecular-weight fatty acids make high melting points.

Procedure

A liquified fat sample is drawn into a capillary tube by capillarity, to a height of about 1 cm. The lower end of the tube is then sealed off in a flame and the tube and contents stored in a refrigerator for 18 hr or longer.

Upon removal from the refrigerator, the capillary tube is secured to a thermometer with a rubber band in such a manner that the sample is adjacent to the bulb of the thermometer. Immerse the thermometer and capillary tube in a tube containing water to a depth of about 3 cm. The latter tube is suspended in a flask or beaker containing water at a height equal to that in the test tube. Heat the water so that the temperature change is approximately 0.5° per min. Record temperature when sample melts, as evidenced by a change from opaque to a transparent appearance.

Measurement of Oxidative Rancidity in Lipids

Rancidity is a term loosely applied to the occurrence of off-flavor and odor in fats. This condition is the result of an oxidation of unsaturated fatty acids; however, hydrolysis, if it occurs, may give rise to off-flavors and odors especially if the fat contains low-molecular-weight fatty acids (butyric, caproic, and caprylic).

Peroxide Value.—Peroxides are the first oxidation products of unsaturated fats and oils. When the concentration of "peroxides" reaches a certain level, complex changes occur with the formation of aldehydes, ketones and hydroxyl groups. The latter products are volatile and are mainly responsible for the off-flavor and odor. Perhaps it should be pointed out that peroxides are not directly responsible for the rancidity; rather the concentration of peroxides is indicative of oxidation during the early stages of the lipid deterioration. The peroxide values is expressed as milliequivalents or peroxide per kilogram of fat.

Materials

(1) 200 g fresh oil. (An oil is recommended rather than a fat because it is easier to handle, being a liquid at room temperature. Do not use olive oil for it reacts with copper to give a green compound that does not catalyze oxidation. Samples should be held in a refrigerator, tightly closed, and away from light.)

(2) 100 g rancid oil. (This can be prepared from the fresh oil by exposing oil to light and air until an off-odor is noted, about 2-3 weeks. To hasten oxidation add 1 ml of alcoholic cupric chloride and bubble air through sample.)

(3) Glacial acetic acid—Chloroform (60:40) ca. 500 ml

(4) Saturated potassium iodide (100 ml)

(5) Soluble starch, 1%

(6) 0.1 *N* sodium thiosulfate, standardized (for preparation and standardization see page 21 of this experiment)

(7) Alcoholic cupric chloride. 0.16 g cupric chloride per 100 ml ethyl alcohol.

Procedure

Weigh four 10 g samples of oil (or fat) to nearest milligram into a 500 ml Erlenmeyer flask (two of the samples fresh, two rancid). Add to each, 50 ml of acetic acid-chloroform and shake until lipid is dissolved. Add 1 ml of saturated potassium iodide solution, rotate the flasks rapidly for 15-20 sec and then allow to stand in the dark for exactly two min. At the end of the period add 100 ml of distilled water to each sample and titrate the liberated iodine with standardized 0.1 *N* sodium thiosulfate. As the endpoint is approached, i.e., when solution becomes a pale yellow, then add several drops of starch solution and titrate to disappearance of blue color.

Run a blank, simultaneously omitting only the oil.

Calculate the peroxide values (P.V.) according to the equation

$$\text{P.V. (meq./kg lipid)} = \frac{(S-B) \times N \times 1000}{W}$$

where
S = ml $Na_2S_2O_3$ to titrate samples
B = ml $Na_2S_2O_3$ to titrate blank
N = normality of $Na_2S_2O_3$ solution
W = weight of sample in grams

Compare peroxide values for fresh oil sample and rancid oil sample.

Detection of Aldehydes.—(Kreis test) Pour about 2 g of the rancid oil into a test tube and a like amount of the fresh sample into another test tube. To each add 2 ml of 1% solution of phloroglucinol in ether and 2 ml of concentrated HCl. Shake tubes. The appearance of a pink color indicates the presence of an aldehyde.

Effect of Pro-oxidants and Antioxidants in Lipids.—Place 15-20 ml of the fresh oil in each of three clean, dry test tubes. Add 1 ml of alcoholic cupric solution to tubes 1 and 2; to tube 2 add 1 ml of 1.5% solution of hydroquinone in alcohol (or any antioxidant such as Tenox, etc.). Tube 3 will serve as a control. Place all three tubes in a gently boiling water bath and pass air through them for 1 hr. At the end of 1 hr test samples for odor, then cool. Weigh a 10 g sample into separate Erlenmeyer flask, and determine peroxide values. Compare values.

Although peroxide number may be used as a measure of lipid quality in the early stages of deterioration, this index becomes less reliable as the degradation of peroxides increases. Additionally, the composition of the lipid will have an effect on the peroxide value during the oxidative process. Triglycerides having relatively high concentrations of polyunsaturated fatty acids will show a high initial peroxide value, while triglycerides with more saturated fatty acids will have low peroxide values. Examples of the former are soybean, cottonseed and salmon oil; of the latter, butter, coconut fat, oleo oil and lard.

Thiobarbituric Acid Value.—The 2-thiobarbituric acid (TBA) test is another method for the measurement of oxidation products of lipids. Thiobarbituric acid reacts with malonyldialdehyde to form a red colored compound (possibly other β-carbonyl products are formed during oxidation). Only fatty acids with two or more double bonds will yield a TBA value. The TBA value is defined as the number of milligrams of malonyldialdehyde per kilogram of sample. (Note: The method is an empirical method and as a consequence must be followed closely to duplicate results. The titration should proceed as rapidly as possible because oxygen will oxide the iodine present.)

Materials

(1) Trichloracetic acid solution. 20 g trichloroacetic acid (TCA) is dissolved in distilled water and made to 100 ml

(2) Pyridine hydrochloride solution. Four parts pyridine is mixed with 7 parts 6 *N* hydrochloric acid. (Some lots of pyridine give high blanks. This may be corrected by refluxing for one hour with 2-thiobarbituric acid at the rate of 1 g/100 ml pyridine, followed by distillation)

(3) 2-thiobarbituric acid (TBA) reagent. One gram of TBA is dissolved in 75 ml 0.1 *N* sodium hydroxide and made to 100 ml with distilled water

(4) Chloroform

(5) HCl-TCA-pyridine reagent. Mix 650 ml of 0.6 *N* hydrochloric acid with 50 ml TCA solution, and add 50 ml pyridine hydrochloride solution

(6) Antioxidant mixture. Eastman Tenox 4 (consisting of 20% butylated hydroxyanisole, 20% butylated hydroxytoluene, and 60% corn oil)

(7) 200 g fresh oil (any oil other than olive oil is satisfactory) store in refrigerator

(8) 100 g oxidized oil (prepared from fresh oil). To hasten oxidation add 1 ml alcoholic cupric chloride, warm, and bubble oxygen through the sample

Procedure

Accurately weigh 100-200 mg of each oil into a 250 ml Erlenmeyer flask. Add 4 ml distilled water, 5 ml pyridine hydrochloride solution, 5 drops antioxidant mixture, 10 ml trichloroacetic acid solution and 6 ml TBA reagent. Avoid shaking the flask. Connect the flask to a water-cooled condenser (West) and lower into a boiling water bath. Reflux exactly 30 min at which time 75 ml of the HCl-TCA-pyridine hydrochloride reagent is added through the top of the condenser. Continue refluxing for 10 min, then cool the flask to room temperature with water. Disconnect the condenser. The color will vary from pink to deep red depending on the extent of oxidation of the sample. Centrifuge about 40 ml of the solution in a 50 ml centrifuge tube for 5 min at 10,000 rpm. Pipet 15 ml of the clear solution into a second centrifuge tube, add 15 ml chloroform, cap, and shake vigorously for about 1 min. Centrifuge the tube for 3 min at 1200 rpm. The colored solution should be free of turbidity. In case of cloudiness, filter the colored solution over anhydrous sodium sulfate. Use a sufficient quantity of the colored solution to read the absorbence at 535 nm.

Run a blank simultaneously, omitting only the oil.

The TBA number is calculated as follows:

$$\text{TBA} = \frac{\text{mg TBA reacting material in sample}}{\text{Sample wt (kg)}}$$

$$= \frac{\dfrac{A \times \text{MW} \times 10^3}{Am\,(100/\text{x})}}{\text{Sample wt (kg)}}$$

$$= \frac{A \times 4.6 \times 10^{-2}}{\text{Sample wt (kg)}}$$

where

A = Absorbence reading

MW = Molecular weight of malonyldialdehyde (72)

Am = Molar absorptivity index of malonyldialdehyde, 1.56×10^5

x = Dilution factor, 10 ml unless further dilution of the sample solution is necessary for an acceptable absorbence

(See Yu, T. C., and Sinnhuber, R. O. 1957. 2-thiobarbituric acid method for measurement of rancidity in fishery products. Food Tech. *32*, 104-108.)

Esterification and Gas-Liquid Chromatography of Neutral Lipids

Lipids are a class of heterogeneous compounds. As a consequence no single solvent is suitable for extracting lipids from naturally occurring materials. Solvents of low polarity, such as carbon tetrachloride, diethyl ether and liquid hydrocarbons, will extract glycerides, sterols and small amounts of complex lipids, whereas use of mixtures of the above solvents and polar solvents, such as alcohol, will extract the majority of complex lipids. Examples of the latter mixtures are ethyl ether/ethanol and chloroform/methanol. Extraction is often performed in a nitrogen atmosphere to prevent oxidation of unsaturated fatty acids of the glyceride.

In this experiment the glycerides and sterols are extracted with ether. The glycerides are separated from the sterols by saponification, the glycerides are esterified, and analyzed by gas-liquid chromatography.

Materials

(1) Diethyl ether

(2) Chloroform

(3) 95% ethyl alcohol

(4) Sodium chloride

(5) 5% HCl in methanol

(6) Petroleum ether, 30-60°

(7) 100/200 mesh Unisil (Carlson Chemical Co., Williamsport, Pa.)

(8) Soxhlet extraction apparatus

(9) Peanut butter

(10) Peanuts or nuts of any kind

(11) Gas Chromatograph

Procedure

Extraction.—Weigh a 1.5-2.0 g sample of peanut butter onto a piece of Whatman No. 1 filter, then fold the filter paper and place in an extraction thimble which, in turn, is placed into a Soxhlet extraction tube. Attach the tube to a weighed flask

(Fig. 4.2). Half-fill the flask with ether and extract the mixture for about 16 hr, warming the flask gently on a heated mantle. Rearrange the apparatus for distillation and evaporate ether to dryness. Redissolve the residue in 5 ml of chloroform.

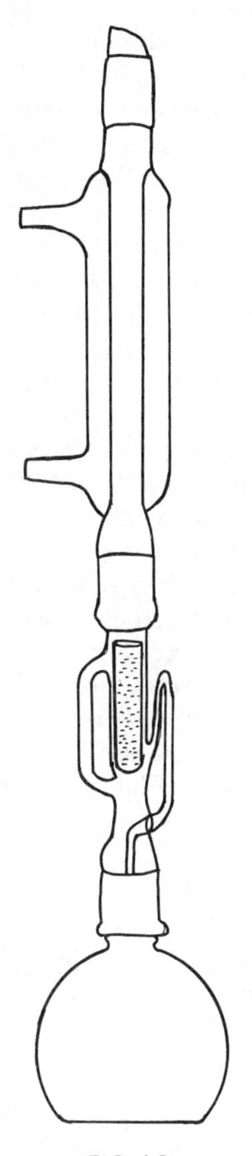

FIG. 4.2.
SOXHLET EXTRACTION APPARATUS

Separation of Neutral Lipids.—A 100/200 mesh Urisil silicic acid column is used to separate neutral lipids from other lipid material (e.g., phospholipids, etc.). Mix 7 g Urisil with a small volume of chloroform and add to a 1 × 15 cm column. Transfer the above lipid material onto the top of the column. The neutral lipids are eluted from the column using approximately 300 ml of chloroform. The neutral lipids are recovered by evaporating the solvent on a rotary evaporator to 10 ml (at room temperature). Transfer to a small flask and evaporate to dryness under a stream of nitrogen. Redissolve the residue in 5 ml chloroform and transfer to a 6 in. test tube. Add 3 ml of 5% HCl in methanol to the flask and cap with a small funnel and a marble. (The funnel and marble act as a crude reflux condenser reducing the amount of alcohol lost.) Place the test tube in a 75° C water bath and allow to reflux for 1 hr. If too much methanol evaporates during reflux, add a little more to the flask. Remove, cool and add an equal volume of water to the test tube. Saturate with sodium chloride and add 2 ml of petroleum ether. Cap, invert and shake. The petroleum ether layer contains the methyl esters of fatty acids.

Gas Chromatography.—Analyze the esters by gas-liquid chromatography to ascertain the composition of the sample. Compute peak areas by triangle technique. Determine the identity of your sample by comparison with standard chromatograms available in the laboratory.

Notes:

(1) A number of reagents are available for esterification of the fatty acids: boron trifluoride-methanol, diazomethane, and HCl-methanol, to mention a few. The method used in this experiment is not the best technique, but it is rapid and utilizes reagents commonly available in the laboratory.

(2) The coating used on the GLC column will determine the position of elution of isomers. Using nonpolar columns such as Apiezon L and SE-30, unsaturated fatty acids elute before their saturated homologs. In contrast, when using polar columns, such as ethylene glycol succinate and diethylene glycol adipate, the unsaturated acids are retained longer than their saturated homologs.

(3) In order to identify a lipid by gas-liquid chromatography certain factors must be known: (a) the identity of the individual fatty acids in the chromatogram (this can be obtained from use of standard fatty acids), and (b) there must be a definite distinguishing factor, i.e., ratio of one acid to another, the presence or absence of a certain acid. Animal fats are relatively high in saturated fatty acids while most vegetable oils are relatively high in monounsaturated and polyunsaturated fatty acids. Vegetable fats have little, if any, fatty acids below myristic (C_{14}) acid. In contrast, butter fat contains fatty acids as short in chain length as butyric (C_4) acid.

QUESTIONS

(1) Why would gas-liquid chromatography analysis be superior to the Reichert-Meissl or the Iodine number in determining the characteristics of a particular fat or oil?

(2) What compounds in the lipid fraction would be classed as nonsaponifiable?

(3) Why are fatty acids converted to esters prior to gas-liquid chromatography?

Proteins

Proteins constitute a major class of food constituents. They are a part of enzymes, antibodies, oxygen carriers, viruses, and other cellular constituents, just to name a few.

The complex chemical composition of proteins makes them quite difficult to characterize by simple chemical or physical procedures. However, their component amino acids may be conveniently detected by various specific chemical tests. Also they may be qualitatively and quantitatively determined by first hydrolyzing the protein, then separating the amino acids by paper, ion-exchange or thin-layer chromatography. Amino acids may also be conveniently determined by converting them to N-acetyl-n-propylesters which are then separated, identified and quantitized by gas chromatography.

PROTEIN TESTS

Qualitative Tests

Egg albumin, lactalbumin or lactoglobulin may be used for the qualitative tests below. The proteins should be essentially free of salts such as chlorides and ammonium salts. The desalting can be accomplished by passing the protein solutions through a sephadex G-25 column or by dialysis. They may then be lyophilized and stored at low temperature (<0° C).

Biuret Reaction.—When a protein is mixed with a solution of sodium hydroxide and a weak solution of copper sulfate, a violet color is produced. This is a test for the peptide linkage and will be positive when two or more peptide linkages are present. The color is due to the presence of a coordination complex with Cu^{2+} in which the four water molecules, normally coordinated with the cupric ion, are displaced by the amino groups. The alkali serves to convert the complex into a soluble sodium salt. The peptide linkage in proteins and peptides can lead to a stable complex with Cu^{2+} in which two five-membered rings are formed. Other compounds also give a positive biuret test; for example, urea forms biuret.

$$NH_2-\overset{\overset{O}{\|}}{C}-NH_2 \overset{\Delta}{\Rightarrow} O{=}C{=}NH + NH_3$$

$$O{=}C{=}NH + NH_2-\underset{\underset{O}{\|}}{C}-NH_2 \Rightarrow H_2N-\overset{\overset{O}{\|}}{C}-\overset{\overset{H}{|}}{N}-\overset{\overset{O}{\|}}{C}-NH_2 \text{ (biuret)}$$

Mix 1 ml of a 2% protein solution with 1 ml of 10% NaOH. Dropwise, add 0.1% $CuSO_4$ solution with mixing. A violet-pink color should develop. Repeat this reaction with a 2% glycine solution. Explain the difference in the reactions.

Millon's Reaction.—When a protein solution is heated with Millon's reagent (solution of mercuric nitrate in nitric acid), a red color is produced. A positive test is due to the presence of a phenolic group (HO-⟨O⟩-) in the protein molecule. Tyrosine is an example of an amino acid that contains this structure.

Place a small amount of powdered protein on a spot plate. Add 5 drops of Millon's reagent. Observe the red color on standing. The test may be repeated using solid tyrosine. To 1 ml of 2% protein solution add 5 drops of Millon's reagent. Heat gently until a red color appears.

Ninhydrin Reaction.—Ninhydrin (triketohydrindene hydrate) reacts with amino acids to produce various shades of blue or purple with the exception of proline and hydroxyproline which give yellow products. The reaction with α-amino acids is:

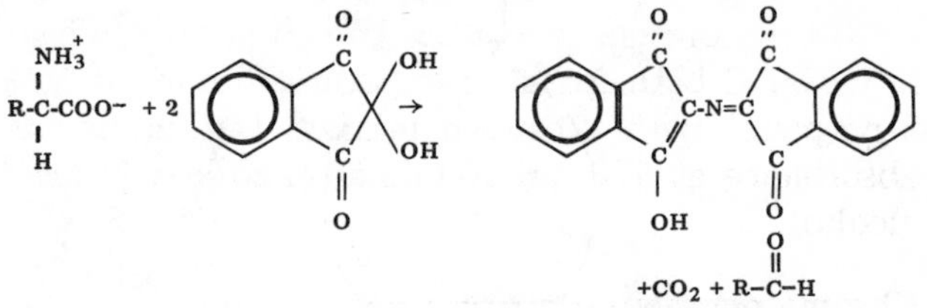

$$R-\underset{\underset{H}{|}}{\overset{\overset{NH_3^+}{|}}{C}}-COO^- + 2\ (\text{ninhydrin}) \rightarrow (\text{product}) + CO_2 + R-\overset{\overset{O}{\|}}{C}-H$$

The blue-purple color may be used to quantitatively determine amino acids and certain peptides by measuring at 570 nm. Commercial automated systems are available for amino acid determinations.

Adjust the pH of a 0.5% solution of protein to pH 7. To 1 ml of this solution add 10 drops of 0.2% ninhydrin solution. Heat in a 100° C bath for 10 min. Note the color.

This test may be repeated using concentrated $(NH_4)_2SO_4$ or partially hydrolyzed protein solutions (Peptone, available from Sigma Chemical Co., St. Louis).

Quantitative Tests

Quantitative Biuret.—Prepare a standard protein solution using crystalline bovine albumin that contains 10 mg per ml. Add 0.25 ml, 0.5 ml and 0.75 ml of this standard to three reaction tubes. Bring the final volume to 1 ml with distilled water. Add 4 ml of biuret reagent to the tubes with mixing. Read the absorbance at the end of 30 min against a blank containing 1 ml of water and 4 ml of biuret reagent.

Concurrently with the standard curve preparation, determine the protein content of an unknown protein solution provided by the instructor. This solution will contain from 5 to 10 mg per ml; therefore, several dilutions should be prepared so that at least two would be in the range of concentration of the albumin standard curve.

Quantitative Ninhydrin.—Using a desalted and lyophilized protein sample, prepare hydrolysis mixture containing 100 mg/ml dissolved in 8 *N* H_2SO_4. Autoclave the sample in an appropriate container for 5 hr at 15 lb.

Neutralize the hydrolyzed sample with solid $Ba(OH)_2$. Barium sulfate precipitates as the $Ba(OH)_2$ is added. Do not add an excessive amount. Neutralize to pH 5. Centrifuge the solution and wash the $BaSO_4$ precipitate with 10 ml of hot water. Save the washes and discard the precipitate. The final volume should be adjusted to 30 ml.

Dilute a portion of the solution 1:100. The final solution, after dilution, will contain 0.033 mg/ml of the original protein. Prepare a standard curve using glycine. Generally a 1 mM (75 μg/ml) solution of glycine may be used. Assay mixtures should be prepared as given in Table 5.1.

After thorough mixing of the components heat in a 100°C bath for 20 min, then add 8 ml of 50% 1-propanol with thorough mixing. Determine the absorbance at 570 nm 10 min after addition of the alcohol.

Chromatography of Protein Hydrolysate

Paper Chromatography.—Use 46 cm × 57 cm Whatman No. 1 chromatography paper. Spot the protein hydrolysate approximately 6 cm from each edge in a corner. Roll the paper and sew the edges together using cotton (colorless) thread. (Paper staples corrode and discolor the paper.) Place approximately 2 cm of the developing solvent (butanol: acetic acid: water - 6:1:2 by volume), in the chromatography tank (30 cm diameter × 60 cm height). Place the cylinders in the tank and allow the solvent front to rise to within approximately 5 cm of the top of the paper. A similar paper can be prepared which contains known amino acids and an unknown hydrolysate spot. When the solvent front migrates the desired distance, remove, mark the solvent front, and air dry. (Be careful not to touch the paper with your fingers—they contain amino acids. Use rubber gloves or forceps). Spray the chromatograms with ninhydrin spray (Dissolve 200 mg of ninhydrin in acetone or commercial aerosol spray cans of ninhydrin are available). Allow to air dry then place in chromatography oven at 90°C until ninhydrin reacts with the amino acids and blue-violet spots appear.

Two-dimensional chromatography may be used if more positive identification is required. In two dimensional chromatography the chromatogram is spotted as before and developed in phenol (chromatography grade liquid) saturated with water. The chromatogram is then dried and washed with ether to remove the phenol. The paper is then turned 90° and the amino acids that migrated on the y-axis are then allowed to migrate along the x-axis using the butanol: acetic acid: water solvent as before. Standards are run for phenol: water as well as butanol: acetic acid: water. These are then compared to the unknown spots. In this manner it is usually possible to positively identify the amino acids of hydrolysates. It should be noted that acid hydrolysis may destroy tryptophan with some losses of serine and threonine.

TABLE 5.1
ASSAY TUBE NUMBER (VOLUME IN ML)

Components	1	2	3	4	5	6	7	8
Standard glycine	0.5	0.25	0.1	—	—	—	—	—
H_2O	—	0.25	0.4	—	0.25	0.4	0.45	2.0
Hydrolysate				0.5	0.25	0.1	0.05	—
Ninhydrin Reagent	1.5	1.5	1.5	1.5	1.5	1.5	1.5	1.5
Caution: Mix and Heat before adding 1-Propanol								
50% 1-Propanol	8.0	8.0	8.0	8.0	8.0	8.0	8.0	8.0

QUESTIONS

(1) What other quantitative procedures are used to determine the concentration of protein solutions? Comment on the advantages and disadvantages of the different methods. (See *Methods in Enzymology*)

(2) What effect does heat treatment have on the solubility and digestibility of food proteins such as milk proteins and bean proteins? (See *Advances in Food Research*)

(3) What methods are used to determine the molecular weight of a protein? What techniques are used to show changes in proteins during processing, e.g., heat treatment?

ISOLATION OF LACTALBUMIN AND GLOBULIN

Isolation Methodology

(1) Preparation of milk serum.—Measure 400 ml of fresh skim milk into a large beaker and heat to 40° C. Add 10% lactic acid solution slowly from a burette with vigorous stirring with a rubber-tipped stirring rod until a flocculent precipitate forms. Allow the precipitate to settle and decant the supernatant fluid through a filter. The casein may be discarded. Heat again to 40° C. Add a few drops of phenolphthalein solution to the filtrate and carefully neutralize to a faint pink with a thin suspension of $Ca(OH)_2$ in water. Filter off the precipitate through a large fluted filter. This precipitate is called the neutralization precipitate of milk serum and consists largely of phosphates. Add dilute (1%) HCl to the filtrate until neutral to litmus. The following tests require 300 ml of filtrate.

(2) Salting out the globulin.—Dilute 50 ml of serum with an equal volume of water. Warm to 30° C and keep at 30° C and saturate the solution with $MgSO_4$ crystals (about 40 g will be required). The precipitate is globulin. All animal globulins are salted out of their neutral solutions on saturation with $MgSO_4$.

(3) Salting out the lactalbumin and globulin.—Dilute 50 ml of serum with equal volume of distilled water, warm to 30° C and saturate with $(NH_4)_2SO_4$ crystals at this temperature (about 80 g of the salt will be required). The precipitate which forms is a mixture of the globulin and the lactalbumin. Compare as well as possible by inspection the relative amounts of the materials salted out in parts (2) and (3).

(4) Separating the globulin from albumin.—Dilute 50 ml of the serum with an equal volume of a saturated $(NH_4)_2SO_4$ solution that has pH 7.0. Let stand for one hour. Filter off the globulin. To the filtrate add 5% H_2SO_4 solution drop by drop with stirring until definite coagulation occurs. This is the albumin.

(5) Heat coagulation of serum proteins.—Dilute 50 ml of serum with an equal volume of water, add 0.3 ml of 10% acetic acid, and heat slowly to 75° C in a water bath. Record the temperature of first cloudiness and the temperature of coagulation. Observe the stages in the heat coagulation of the proteins in this test, i.e., (a) the denaturing as shown by the cloudiness, (b) the agglutination of the denatured colloidal particles. Now raise the temperature of the solution to boiling and boil for 5 min. Filter off the coagulated material while the solution is still hot, in order to facilitate filtration. Test the filtrate for protein, using the acetic acid-ferrocyanide test, as follows: To 10 ml of water-clear filtrate add 1 ml of 10% acetic acid, and then add, drop by drop, with shaking after each addition, a 5% solution of potassium ferrocyanide. A positive test for protein is shown by a distinct cloudiness.

(6) Desalt by Sephadex or dialysis and lyophilize.

QUESTIONS

(1) In Part 1 of isolation and methodology for lactalbumin and globulin, why did the serum proteins not precipitate with the casein since both are essentially the same isoelectric pH?

(2) Why was $Ca(OH)_2$ added to the whey?

(3) What quantitative relationship exists between the ionic strength of a salt solution and the solubility of a protein in this solution?

PROTEIN AND AMINO ACID CONTENT OF SWEET POTATO

In vivo assays are generally the methods of last recourse for the evaluation of protein quality because animal tests are both time- and material-consuming. Furthermore, the results are open to a wide range of interpretations, yielding valid information only in the hands of the expert. For these reasons *in vitro* methods, be they chemical, microbiological or enzymic, are the methods of choice and when used with discrimination serve as an indispensible compliment to animal assays.

The chemical score method is a reasonably successful method for the evaluation of proteins. It is based on amino acid content and the results are expressed as a percentage value of the ratio of the quatity of each essential amino acid in a test protein to the quantity of each essential amino reference protein. Obviously, the choice of the reference protein will have an affect on the score of a test protein. Standard reference proteins includes egg, human milk, cow's milk and the FAO reference pattern of amino acid requirements (1957).

Procedure

Protein and dry matter.—Samples for analyses are obtained by cutting 3 mm diameter plugs from the mid section (equator) of 3 to 5 sweet potatoes with a cork borer. Kjeldahl N is determined on 4-6 of the plugs (weighed to nearest 0.1 mg) and reported as protein after multiplying by 6.25. Dry matter is determined by drying 6-8 g samples in a vacuum oven at 60°C for 16-18 hr.

Extraction of protein.—6-8 sweet potatoes are peeled and diced, and 300 g are blended in a Waring blender with 600 ml cold water for 4 min at high speed. Filter the resulting slurry through pelon to remove cell walls and fibers. Most of the starch granules are sedimented by centrifugation at 300 × g for 25 min at 4°C. The starch pellets are resuspended and centrifuged at 16,000 × g for 10 min. The centrifugal supernatants of each sample are combined and proteins then coagulated by adding trichloracetic acid to 12% concentration and heating to 50°C. The coagulum was precipitated at 10,000 × g for 25 min, then extracted with a mixture of acetone-ether (1:1) until extracts are colorless. Dry the protein powders overnight at 18-20°C in a hood and store in an evacuated desiccator. Kjeldahl N analyses are run on the residue.

Hydrolysis of Protein.—Weigh about 60 mg of the above protein powder into 10 ml ampoules, then add 8 ml 6 *N* HCl. Ampoules are degassed with nitrogen and sealed under 0.5 atm of nitrogen. Place sealed ampoules in a reaction vessel and heat with refluxing toluene for 16 hr. Transfer the contents of the ampoule to centrifuge tubes and centrifuge at 15,000 × g for 10 min. Pour the supernatant liquid into a 20 ml beaker, wash the ampoule with 2 ml distilled water and add to supernatant liquid. Dry sample over sodium hydroxide pellets in an evacuated desiccator. The dried samples are then dissolved in 5 ml sample dilution buffer (pH 2.2)[1] and filtered through a 0.2 micron millipore filter.

Since tryptophan is known to be destroyed by acid hydrolysis, another set of samples must be weighed and hydrolyzed as before except 0.17 ml thioglycholic acid is added.

Ion Exchange Cleanup.—The hydrolysate is highly colored due to the presence of sugars. These interfering substances can be removed through the use of an ion exchange column. Dowex 50 W, 200-400 mesh, is prepared by three alternate washings with 1 *N* HCl and water, then dried at room temperature overnight. Columns are prepared by suspending 0.35 g of the above prepared resin in 2 ml 1 *N* HCl and transferring the suspensions into a Pasteur pipet (Fisher Cat. #13-678-5A). The acid (*N* HCl) is allowed to drain until it reaches the top of the resin, then add 1 ml water and allow to drain to the top of the column; *at no time do you allow the top of the column to become dry.* Transfer the hydrolysate to the column; when level of hydrolysate nears top of the resin column wash three times with 1 ml portions of 0.5 *N* acetic acid followed by two 1 ml portions of water. The amino acids are eluted with 4 ml of an eluant containing 20% triethylamine, 10% acetone and 70% water. The eluate is then dried over sulfuric acid in a partially evacuated desiccator (Note: Care should be taken to prevent bumping when applying vacuum to the desiccator).

After drying, take up in 5 ml "sample dilution buffer" for the amino acid analyzed.

Amino Acid Analysis.—The mixture of amino acids is resolved according to the method of Sparkman et al. (1958) using either a Beckman Model 119 or a Jelco Model JLC-5AH amino acid analyzer. Norleucine is added to the sample dilution buffer as an internal standard for the long column and beta-guanidine-propionic acid for the short column.

The amino acids are measured as micro moles per sample and reported as moles per 100 kg based upon the Kjeldahl protein content of the powder. In essence the method is as follows: A mixture of amino acids is applied to an ion-exchange resin. Buffers of carefully controlled pH and ionic strength are pumped through the ion-exchange resin column, where the individual amino acids are selectively bound and then eluted on the basis of the chemical nature of the resin, the temperature, and the pH and ionic strength of the buffer used. The resolved components of the hydrolysate, after being pumped out the bottom of the column, are monitored by mixing the amino acids with ninhydrin, reacting the mixture in a heated coil then passing the stream of colored substances through a colorimeter. A photocell at the cuvettes transmits the light response (570 nm and 440 nm) to a recorder which prints out the chromatogram. The chromatogram is then interpreted on the basis of previously accumulated data, i.e., calibration runs with known amino acids under operating conditions exactly the same as the above run.

---

[1] Sample dilution buffer (pH 2.2)—Sodium citrate. $2H_2O$ 19.6 g, conc. HCl 16.5 ml, Thiodiglycol (25% solution)—10.0 ml and caprylic acid—0.1 ml made up to a final volume of 1 liter.

Enzymes

IMMOBILIZED ENZYMES

The basic purpose for the immobilization of enzymes is to retain the enzyme within prescribed physical boundaries while at the same time allowing substrate free access through the enzyme region. Hence the enzyme phase is easily separable from the substrate and product and, in fact, systems can be designed whereby this separation can occur on a continuous basis. From the inception of the immobilized enzyme concept, it was realized that this inherent ability to confine an enzyme to a given physical region and to reuse the enzyme constituted a tremendous financial and bioengineering advantage.

There are two general schemes for immobilizing an enzyme: 1) entrapment and 2) attachment to a surface. Entrapment consists of surrounding volumes of an aqueous enzyme solution by semipermeable physical boundaries, e.g., entrapment within hollow fibers, gels, and fibers; microencapsulation; and entrapment behind ultrafiltration membranes. Attachment to a surface can be achieved by ionic binding, e.g., ion exchange resin; or by covalent bonding to a surface, e.g., synthetic membranes, organic porous polymeric particles and inorganic porous glass or ceramic particles. Attachment differs from entrapment in that: 1) the enzyme's conformation may be modified by the attachment procedure, 2) the enzyme is insolubilized, and 3) the enzyme is subjected to the environment established by the surface to which it is attached.

The properties of an immobilized enzyme system can be described by: 1) a change in enzyme conformation, 2) a different environment than in solution, 3) a partitioning of substrate between enzyme and liquid phases, and 4) diffusion limitation. In general, most immobilized enzyme systems experience diffusion limitation wherein the flux of substrate into the enzyme region is less than the potential catalytic rate and thus the catalytic rate is controlled by the substrate's diffusion rate into the region. The first experiment using an entrapment method explores the phenomenon of difusion limitation. The second experiment, using ionic binding, explores the effect of matrix charge on the pH profile of the enzyme.

Entrapment

Stock solutions:

Trypsin: 1 mg/ml in 0.001 *M* HCl

Substrate, TAME (p-toluenesulfonyl-L-arginine methyl ester, M.W. 378.87): 0.01 *M* in distilled water.

Buffer: Tris, 0.046 *M*, pH 8.1 containing 0.0115 *M* $CaCl_2$

Procedure

Free Solution Enzyme.—Dilute the stock trypsin solution to 0.01 mg/ml with Tris buffer. Pipet 0.3 ml of the substrate solution and 2.7 ml buffer into a 1 cm cuvette. Mix and "zero" spectrophotometer at 247 nm. Into a second cuvette pipet 0.3 ml substrate, 2.6 ml buffer and initiate the reaction (zero time) with the addition of 0.1 ml of the diluted enzyme. Mix and record absorbancy at 30 sec intervals for 5 min. Plot absorbancy versus time, and determine the slope of the linear portion of the curve, i.e., ΔA/minute. This is a measure of the zero order rate at the enzyme concentration used.

Entrapped Enzyme.—Label 10 test tubes consecutively. Into a 250 ml beaker, place 87 ml of the Tris buffer and 10 ml of the substrate solution. Place beaker on a magnetic stirrer. Knot and tie with dental floss, one end of an 8 in. length of wet dialysis tubing. Place a 1 in. Teflon stirring bar into the tubing and pipet 3 ml of the dilute enzyme solution into the tubing. Knot and tie the tubing so as to keep the tubing containing the enzyme solution to a minimum length. Remove the excess tubing and dental floss with a pair of scissors. Rinse the external surface of the tubing sac with distilled water to remove any enzyme. To initiate the reaction (zero time) place the entrapped enzyme into the beaker and turn the stirrer "on." Take 3 ml aliquots at 30 sec intervals from the time the enzyme is introduced and place in the labeled test tubes. Sample over a total time interval of 5 min. Determine the absorbency of the aliquots at 247 nm. To compensate for the change in volume in the system the absorbency must be multiplied by the fraction of the original volume at the time the respective aliquot was taken. The proper factor to use is given in Table 6.1. Plot the

TABLE 6.1
FACTORS FOR VOLUME COMPENSATION

	Test Tube No.									
	1	2	3	4	5	6	7	8	9	10
Time (min)	0.5	1.0	1.5	2.0	2.5	3.0	3.5	4.0	4.5	5.0
Factor	1.0	0.97	0.94	0.91	0.88	0.85	0.82	0.79	0.76	0.73

adjusted absorbencies versus time and determine the slope of the linear portion of the curve. Compare this rate to the free solution rate. Since the total enzyme concentration in each system is the same, then by dividing the entrapped enzyme rate by the free solution rate one can obtain the fraction of enzyme actually utilized in the entrapped enzyme system.

Attachment by Ionic Binding

Stock solutions:

Buffer A: 0.02 *M*, Tris, pH 7.0, containing 0.012 *M* $CaCl_2$

Buffer B: 0.01 *M*, Tris, containing 0.012 *M* $CaCl_2$, pH adjusted to various values

Trypsin: 0.5% in buffer A

Substrate: A—0.001*M* TAME, pH 7.0, Buffer B
B—0.001*M* TAME, pH 8.0, Buffer B
C—0.001*M* TAME, pH 9.0, Buffer B
D—0.001*M* TAME, pH 10.0, Buffer B

(Note: Trypsin and substrate solutions should be made just prior to the experiment to minimize autolysis and spontaneous hydrolysis, respectively.)

Procedure

To 0.5 g of CM-cellulose in a 100 ml beaker add 100 ml of Buffer A, stir, let settle for 15 min, then decant and discard the cloudy supernatant. Repeat with another 100 ml of Buffer A. Resuspend in a small volume of Buffer A, transfer to a 20 ml beaker and decant off excess liquid. Add 10 ml of the 0.5% trypsin solution and stir with a magnetic stirring bar for 30-60 min. Transfer the mixture to a filter and wash with 300 ml of Buffer A in 50 ml aliquots. Insert a small plug of glass wool into a Pasteur pipet and clamp the pipet to a ring stand. Remove the enzyme-cellulose from the filter with Buffer A and place in a 100 ml beaker. Transfer part of this slurry to the pipet column to obtain a column height of 2-3 cm. Do not allow the column to run dry (the column can be stoppered with a rubber Pasteur pipet bulb). Run each of the four substrate solutions over the column while maintaining a fairly constant liquid level above the bed of the column. Discard the first 20 ml of effluent of each substrate solution and collect the next 5 ml in a labeled test tube. Determine the absorbency at 247 nm using the respective substrate stock soluton as a blank. Record and compare the absorbencies to determine where the maximum reaction occurred (in free solution trypsin has a pH optimum at pH 8.1 for TAME). If the absorbencies are greater than 0.3 Å or too low to read, decrease or increase the column height, respectively. Since the CM-cellulose has a negative charge, we would expect an increase of hydrogen ions within the matrix causing a lower pH than in the bulk solution. This should shift the pH optimum to a more alkaline value.

QUESTIONS

(1) What methods other than those presented are available for immobilizing enzymes?

(2) Do immobilized enzymes show the same kinetic parameters (e.g., Km, Ks, Vmax) as native soluble enzymes?

(3) What industrial uses are being made of immobilized enzymes?

YEAST INVERTASE: ISOLATION AND PURIFICATION BY GEL CHROMATOGRAPHY

Yeast invertase catalyzes the conversion of sucrose to fructose and glucose.

The use of invert sugar (the product of invertase action) is quite widespread in the food industry. Although the hydrolysis can be carried out by acid catalysis, enzymes are much more rapid in their action. This experiment is designed to illustrate enzyme isolation and purification and to study specific properties of the enzyme invertase. Also, the technique of gel chromatography is illustrated.

Procedure

Reagents and Supplies:

(1) 50 ml chromatography columns (Note: 50 ml g.s. burettes may be used)

(2) Sephadex G-200 (medium or coarse, fine does not allow sufficiently fast elution rate) Treat and pack column according to manufacturer[1]

[1] Pharmacia Fine Chemicals, Inc., Piscataway, N.J.

(3) 0.1 M acetate buffer pH 4.8 prepared by adding 0.1 moles of acetic acid (6 g) to approximately 900 ml of water, adjusting the pH to 4.8 with dilute NaOH, and diluting to 1000 ml. Recheck pH after dilution

(4) Fleischmann's Active Dry Yeast or equivalent

(5) 0.1 M $NaHCO_3$

(6) 3,5-Dinitrosalicylate Reagent. Prepared by dissolving 10 g of 3,5-dinitrosalycylic acid in 400 ml of 1 N NaOH. This solution is then mixed with 500 ml of water containing 300 g of sodium potassium tartrate tetrahydrate. The solution is then made up to 1 liter.

Experimental.—Active dry yeast is mixed with 0.1 M $NaHCO_3$ (3-4 ml of $NaHCO_3$ per g of dried yeast) and a smooth paste prepared. The paste is incubated at 35-40°C for 24 hr. After incubation centrifuge at 15,000 × g and retain the supernatant.

The Sephadex G-200 (treated according to manufacturer) is poured into the column to give a bed volume of approximately 20-30 ml. Do not allow the level of the buffer to drain below the top of the Sephadex bed. Allow approximately 1 ml of buffer to remain above the top of the gel bed.

Place 2-4 ml of yeast extract on the column and allow to flow into the Sephadex followed by elution with 0.1 M acetate buffer. This generally requires about 1.5-2.5 hr to elute the enzyme and the bulk of protein which elutes after the enzyme. The inert protein that elutes after the enzyme has a yellow color and may be used to ascertain when the enzyme has eluted.

Fractions should be collected at 2 ml intervals. The enzyme can usually be isolated in the first 30-40 ml. An automatic fraction collector will simplify the collection but is not essential. After collecting, enzyme activity and enzyme protein concentration are determined. All assays can be carried out at approximately room temperature.

The enzyme activity is determined by measuring the amount of reducing sugars (glucose and fructose) formed by incubating the enzyme with sucrose. Appropriate concentrations of the enzyme (0.1 ml of 1:1, 1:100 and 1:000) are incubated with 300 μ moles of sucrose in 5 ml of 0.05 M acetate buffer, pH 4.8. After 10 min of incubation (for extremely active preparation this time may be shortened to as low as 1 min) add 1 ml of 3,5-dinitrosalicylate reagent to stop the reaction. The reaction mixture is diluted to 20 ml with water and the absorbence measured at 540 nm. A standard curve should be prepared in the range of 0 to 300 μ moles of glucose. Be sure that the standard curve is linear. If the incubated enzyme samples show greater absorbence than the highest standard, dilution greater than 1:1000 may be required.

The protein analysis may be performed by a number of methods, including biuret (see Proteins experiment) or by direct spectrometric methods, e.g.,

$$1.74 \times A_{280} - 0.45 \times A_{260} = \text{mg of protein/ml}$$

After determining the protein concentration, calculate the specific activity of the enzyme. Specific activity is defined as the number of μmoles of sucrose hydrolyzed per minute (units) per milligram of protein. High specific activity is desired since this would indicate a minimum amount of inert protein.

Additional experiments may be performed on the enzyme preparation. For example, incubate an appropriate dilution of the enzyme with various concentrations of sucrose and plot the activity versus the concentration of sucrose. This will permit the construction of the classical Michaelis-Menten (Lineweaver-Burk modification) graph and the subsequent Michaelis-Menten constant (Km). Also the maximum velocity (V_{max}) may be calculated.

Other experiments can relate to the effect of pH on activity and the actual breakdown of sucrose in various foods. Also the inhibition of p-mercuribenzoate (approximately 10^{-4} to $10^{-5}M$) and glucose may be studied.

QUESTIONS

(1) What is the purpose of determining both the protein and the activity of the enzyme?

(2) What are Sephadex and similar gels composed of and how do they function?

(3) Why do you suppose that the yeast contains invertase?

POLYPHENOL OXIDASE [TYROSINASE, CATHECHOLASE] *O*-DIPHENOL: O_2 OXIDOREDUCTASE; EC 1.10.3.1

Browning of fruits such as apples, pears, peaches and apricots, and vegetables such as potatoes, occurs when the tissue is exposed to oxygen. The exposure is commonly due to bruises, cuts and other injury to the peel. The browning reaction occurs when the enzyme-catalyzed reaction of oxygen and certain phenolic compounds produces quinone structures and their polymerization products which are responsible for the brown color. The enzyme that catalyzes these reactions is more frequently called polyphenol oxidase (*O*-diphenol: oxygen oxidoreductase; EC 1.10.3.1). This enzyme has also been referred to as phenolase, tryosinase, catechol oxidase and potato oxidase. These names are taken from the varied substrate specificity of the enzyme or its source.

Polyphenol oxidase is a copper-containing enzyme which can undergo reversible oxidation and reduction in the process of hydroxylation and oxidation. In hydroxylation, Cu^+ is oxidized to Cu^{2+} and in oxidation, Cu^{2+} is reduced to Cu^+. Since mushroom polyphenol oxidase can exist in multiple molecular form, it can exhibit variable hydroxylation and oxidation activity depending on the multiple form present.

The reaction catalyzed in this experiment is:

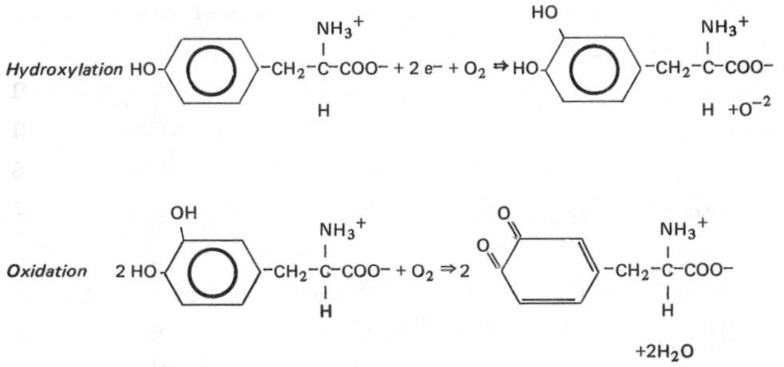

Hydroxylation HO-C6H4-CH_2-CH(NH_3^+)-COO$^-$ + 2 e$^-$ + O_2 ⇒ HO(HO)-C6H3-CH_2-CH(NH_3^+)-COO$^-$ + O^{-2}

Oxidation 2 HO(OH)-C6H3-CH_2-CH(NH_3^+)-COO$^-$ + O_2 ⇒ 2 O=(O=)C6H3-CH_2-CH(NH_3^+)-COO$^-$ + $2H_2O$

Method

Polyphenol oxidase hydroxylates tyrosine to *o*-dihydroxy-phenylalanine (DOPA). *o*-Quinone DOPA absorbs strongly at 280 nm which serves as the assay method. The rate of increase in absorbence at 280 nm is proportional to the enzyme concentration and is linear for approximately 5 to 10 min after the initial lag.

Reagents:

Enzyme: Aqueous solution containing 0.25-0.50 mg per ml

Substrate: 1 mM L-tyrosine (aqueous solution)

Buffer: 0.5 *M* phosphate buffer, pH 6.5

NaCL: 1.0 *M* (aqueous solution)

$NaHSO_3$: 67 μM (aqueous solution)

Experimental Procedure.—Using 3 ml silica cuvettes prepare assay mixtures as given in Table 6.2.

Add all of the components except the enzyme. After mixing, bubble oxygen into the mixture for 5 min. The reaction is initiated by adding 0.1 ml of enzyme followed by thorough mixing. A recorder set for 1.0 Å full scale is suitable. The activity of the enzyme is determined by calculating ΔA per min for the initial velocity. One unit of activity is commonly defined as the amount of enzyme required to give 0.001 Å change per min. The enzyme should be diluted to give approximately 0.1 Å change per min in the 3 ml of reaction mixture.

A Lineweaver-Burk plot may be prepared by determining the activity at several (usually seven) substrate concentrations. Velocity^{-1} is plotted on the Y-axis versus [Substrate]$^{-1}$ on the X-axis. The type and degree of inhibition may also be determined by using Lineweaver-Burk or Dixon plots.

In the Dixon plot, velocity^{-1} is plotted versus [Inhibitor] using at least two substrate concentrations.

QUESTIONS

(1) Calculate the number of units of enzyme used in the assays.

(2) Calculate the percent inhibition for the various inhibitors.

(3) Why was the reaction mixture saturated with oxygen?

TABLE 6.2
ASSAY MIXTURES FOR POLYPHENOL OXIDASE EXPERIMENT

	Assay Number (Volume in ml)						
Components	1	2	3	4	5	6	7
Buffer	1.0	1.0	1.0	1.0	1.0	1.0	1.0
Substrate	1.0	1.0	1.0	1.0	1.0	1.0	1.0
H_2O	1.0	0.95	0.9	0.5	—	0.5	—
Enzyme	—	0.05	0.1	0.1	0.1	0.1	0.1
NaCL	—	—	—	0.4	0.9	—	—
$NaHSO_3$	—	—	—	—	—	0.4	0.9

Vitamins

DETERMINATION OF ASCORBIC ACID (VITAMIN C) IN ENRICHED PUDDING

Ascorbic acid is very susceptible to oxidation. Its destruction is accelerated by alkalies, iron and copper salts, heat, oxidative enzymes, air and light. It is readily preserved in acid media, but it disappears rapidly when heated in neutral and alkaline media. Certain respiratory enzymes destroy ascorbic acid and, as a consequence, loss of vitamin C during fresh storage of fruits and vegetables may be considerable. Blanching destroys enzymes and tends to preserve the vitamin content of processed foods. L-ascorbic acid is the physiologically active compound. It readily loses two of its hydrogen atoms to form L-dehydroascorbic acid. Reactions between these two forms are reversible but further oxidation results in a loss of vitamin activity. Thus, owing to the chemical nature of ascorbic acid, ascorbic acid is frequently assayed in foods.

Procedure

Reagents and Apparatus

(1) *o*-Phenylenediamine solution—For each 100 ml of solution required, weigh 20 mg *o*-phenylenediamine dihydrochloride (Eastman Organic Chemicals Department, Distillation Products Industries, Rochester, New York, 14603, No. 678). Dilute to volume with H_2O immediately before use.

(2) Ascorbic acid Standard Solution—Weigh accurately 25 mg USP Reference Standard Ascorbic Acid (USP Reference Standards, 46 Park Avenue, New York, New York, 10016) that has been stored in a desiccator, out of direct sunlight. Transfer to a 100 ml volumetric flask. Dilute to volume immediately before use with metaphosphoric acid-acetic acid extracting solution(5). Dilute 3 ml to 100 ml with extracting solution (1 ml = 7.5 μg ascorbic acid). Designate as Standard Solution and proceed as in Determination.

(3) Sodium acetate solution—Dissolve 500 g $NaOAc.3H_2O$ in H_2O and dilute to 1 liter.

(4) Boric acid-sodium acetate solution—Dissolve 3 g H_3BO_3 in 100 ml sodium acetate solution (3). Prepare fresh solution for each assay.

(5) Extracting solution—Metaphosphoric acid-acetic acid extracting solution. Dissolve, with shaking, 30 g HPO_3 pellets in 80 ml glacial HOAc and about 500 ml H_2O. Dilute to 1 liter.

(6) Acid-washed Norit—Add 1 liter 10% HCl to 200 mg Norit. Heat to boiling. Filter with suction. Remove cake of Norit to large beaker. Add 1 liter H_2O, stir and filter. Repeat washing with H_2O and filtering. Dry overnight at 110-120°C.

(7) Celite—Filter aid.

(8) Vortex mixer—Scientific Industries, Inc., or equivalent.

(9) Fluorometer—Farrand fluorometer, Model A-3 equipped with the filters used for vitamin B_1 assay (thiochrome measurements) or equivalent instrument.

(10) Waring blender or equivalent.

Sample Preparation.—Weigh 100 g sample into blender jar, add 400 ml metaphosphoric acid-acetic acid extracting solution (5), and blend very briefly until sample is dispersed. Centrifuge an aliquot of ca. 100 ml. Designate as Sample Assay Solution, and proceed to Determination.

Determination.—The following steps must be performed consecutively without delay.

Transfer 40 ml of Standard and Sample Assay Solutions to separate 50 ml glass-stoppered centrifuge tubes. Add 0.5 g acid washed Norit (6), shake vigorously, add 2 g Celite, shake briefly, and centrifuge at 2000 *g* for 10 min. (Solution should be clear. Otherwise, decant almost clear solution into another centrifuge tube, add 2 g of Celite, and centrifuge again.)

Transfer 10 ml of each filtrate to 25 ml volumetric flask containing 5 ml of boric acid-sodium acetate solution (4). Let stand 15 min, swirling occasionally. Designate as standard or sample blank solutions.

During the 15 min period, transfer 5 ml of each filtrate to a 25 ml volumetric flask to which has been added 5 ml sodium acetate solution (3) and dilute to volume with water. Transfer 5 ml of each solution to each of three test tubes. Designate as standard or sample tubes.

At appropriate time, dilute the blank solutions to volume with water. Transfer 5 ml of these solutions to each of three test tubes. Designate as standard or sample blank tubes.

Add 5 ml *o*-phenylenediamine solution (1) to all tubes. Use Vortex mixer to swirl tubes. Protect from light and let stand 35 min at room temperature. Transfer solutions to fluorescence reading tubes.

Fluorometry Measurement.—Measure fluorescence of standard tube; call this reading *A*. Next, measure fluorescence of standard blank tube; call this reading *B*. Then measure fluorescence of sample tube; call this reading *C*. Finally, measure fluorescence of sample blank tube; call this reading *D*. Calculate as follows:

Ascorbic Acid in mg/5 oz =

$$\frac{(\text{Avg } C - \text{Avg } D)}{(\text{Avg } A - \text{Avg } B)} \times 7.5 \times \frac{500}{100} \times \frac{28.35}{1000} \times 5 = \frac{(\text{Avg } C - \text{Avg } D)}{(\text{Avg } A - \text{Avg } B)} \times 5.32$$

DETERMINATION OF VITAMIN A

There are a number of methods for the analysis of Vitamin A in foods, e.g., rat bioassay, chemical methods and physical methods. These analyses have been combined with a number of preparatory steps in order to improve accuracy. This is especially true for foods because of the presence of interfering substances such as Vitamin A esters, Vitamin A isomers, carotenoids and Vitamin A decomposition products. These difficulties compel the use of saponification solvent extraction and, in some cases, chromatography on alumina and magnesia to prepare the samples for analysis.

Ultraviolet Absorption

Ultraviolet absorption is one of the most common methods for measuring Vitamin A activity (other than the Carr-Price Method) because the absorbence value of a properly prepared solution at 325 nm is directly proportional to Vitamin A concentration. The absorption value can then be converted to gravimetric units by means of a conversion factor.

Procedure

Weigh by difference, 50 g of Vitamin D homogenized milk and add into a saponification flask equipped with a standard taper joint. Add 50 ml of alcoholic potassium hydroxide solution, prepared by dissolving 12 g of KOH in 100 ml methanol, and attach to a reflux condenser. Heat for 15 min under a nitrogen atomosphere or until saponification is complete. The solution is then allowed to cool to room temperature without external cooling. Wash the condenser with 10 ml of water. Transfer contents of the saponification flask to a separatory funnel. Wash the saponification flask with 100 ml acetone-hexane (1:1), or more if necessary to equal the volume of alcoholic potassium hydroxide used. Add the solvent mixture to the separatory funnel and allow the layers to separate; then drain the aqueous layer into another separatory funnel. Repeat extraction of aqueous layer twice with 75-100 ml of hexane. Shake moderately. Combine all solvent extracts and wash with several portions of water until free of alkali, testing with phenolphthalein (test paper). The hexane extracts are then dried over ca. 20 g anhydrous sodium sulfate. Filter the hexane extracts through anhydrous sodium sulfate, placed on a plug of Pyrex glass wool in a funnel, into a 500 ml round bottom flask. Rinse the separatory funnel with two 10 ml portions of hexane and add the rinses to the 500 ml flask, through the funnel. The hexane extracts are evaporated on a water bath (30°C) to a low volume (ca. 25 ml), under a nitrogen atmosphere, then transfer to 100 ml round bottom flask and continue evaporation to dryness. Take up the residue immediately in 10 ml chloroform. (Add to the chloroform solution 4 ml of standard saponified Vitamin A solution, then dilute to volume. If the unknown contains only a small amount of Vitamin A, fortify with the Vitamin A before the final evaporation step.)

Liquid Chromatography

Liquid chromatography presents a fast and reliable method for the quantitative analysis of Vitamin A. The fat-soluble vitamins are best chromatographed by the reversed-phase technique wherein a polar water-alcohol solvent is used with a nonpolar stationary phase. In this kind of chromatography compounds usually elute in order of their decreasing water solubility. The chromatographic conditions required for single component analysis of Vitamin A are summarized below:

Instrument	DuPont 820 LC	Varian Aerograph 4100
Column	Permaphase ODS	Micropak 50 cm
Mobile Phase	95% Methanol/5% water	Pump A — hexane; Pump B — 90% methylene chloride + 10% 2-propanol. Set programmer. Final extra ratio 81% Solution (A) and 19% Solution (B)
Column Temperature	50°C	21°C
Column Pressure	1200 psi	1200 psi
Flow Rate	2 ml/min	5 ml/min

Detector	UV Photometer (325 nm)	UV Photometer
Sample Size	50 microliters	50 microliters
Retention Time	40 sec	10 min

Vitamin A Standard.—The U.S. Pharmacopaeia reference standard consists of 3.44 mg Vitamin A acetate (equivalent to 3.00 mg vitamin alcohol) per gram of cottonseed oil. The potency per gram assigned to this is 10,000 USP units.

An accurately weighed 50 mg of crystallized Vitamin A acetate is made up to 50 ml with chloroform (this is designated stock solution). One ml of the stock solution is diluted to 100 ml with chloroform. The following volumes of the latter diluted solution are placed in a 1.0 cm cell and read at 325 nm. (Dilute each to 1.0 ml total volume.)

0.80 ml = 8 μg Vitamin A acetate
0.60 ml = 6 μg Vitamin A acetate
0.40 ml = 4 μg Vitamin A acetate
0.20 ml = 2 μg Vitamin A acetate

The measured absorbence values are plotted against the amount of Vitamin A acetate.

An Alternative Method

Express the contents of one capsule of U.S.P. Vitamin A Reference Standard into a 300 ml standard taper ground glass Erlenmeyer flask. (Each capsule contains approximately 250 mg of cottonseed which is standardized to contain in each gram 34.4 mg of all-trans retinyl acetate equivalent to 300 mg of retinol.) Add 25 ml H_2O and 25 ml of methanol saturated with potassium hydroxide and a small spatula tip of EDTA. Reflux for 15 min and cool in an atmosphere of nitrogen.

Extract once with acetone:hexane (1:1), reextract the aqueous layer twice with hexane. Combine all extracts and wash free of alkali with water, testing with phenolphthalein test paper (or litmus paper). Dry with anhydrous sodium sulfate, then evaporate and make to volume of 100 ml (stock solution).

Dilute 5 ml of stock solution to 50 ml with hexane and read at 325 nm with a spectrophotometer. The following equation can be used.

$$\frac{\text{A @ 325 nm}}{0.182} = \mu\text{g/ml or mg/L Vitamin A alcohol}$$

For example, standard reading usually is 1.10.

$$\text{ca. } \frac{1.10}{0.182} = 6.04\ \mu\text{g/ml}$$

Take 4 ml of the diluted Vitamin A solution and make to 10 ml with chloroform. Inject 50 μl of the latter solution into the LC column and measure. Compare with an unknown sample.

Chromatography Programer Schedule

(1) Initial: 19% B (90% methylene chloride + 10% 2-propanol); 81% A (hexane). (2) Decrease Pump B 9% per min for 2 min (Pump A is then 99%). (3) Hold Pump B at 1% for 1 min. (4) Decrease Pump B to 0.2% for 1 min (Pump A is then 99.8%). (5) Decrease Pump B to 0.08% for 1 min. (6) Reset Pump B to 19%.

QUESTIONS

(1) What general advantages and disadvantages does fluorescence offer as compared to ultraviolet or visible absorption techniques?

(2) Could vitamin A be analyzed by gas-liquid chromatography?

(3) Would vitamin A precursors such as β-carotene be analyzed by the same method used for vitamin A?

Metal Analysis in Foods Using Atomic Absorption

The classic methods for determining the major metals and anions, which include generally a digestion and/or ashing of the foodstuff followed by a gravimetric, titrimetric, or colorimetric analysis, have generally been sufficient for elements in rather large quantities, e.g., calcium, magnesium, phosphates, and sulfates. However, to actually determine small differences in concentration of various metals or, even more importantly, the presence and concentration of the trace minerals, the classical methods are not usually sufficiently sensitive to yield such information.

Atomic absorption spectrophotometry provides one of the most useful and convenient means for the determination of metallic elements in solution from a wide variety of samples. The sample must be solubilized in aqueous or other solvents in order that it can be aspirated into the flame of the atomic absorption spectrophotometer. An alternative to this solubilization process is the use of the graphite or carbon rod furnace which is gaining wide acceptance for solid sampling.

The analysis of the metallic elements in foods such as beverages, e.g., juices, beer, tea, coffee, etc., is easily accomplished by direct aspiration into the atomic absorption flame or direct injection into the graphite furnace. However, foods such as animal and plant tissues, and fluids with considerable solids must be treated to either decompose the solids or extract the metals. Ashing and/or wet digestion are the most commonly used methods. Ashing is usually performed at temperatures less than 500°C and digestion is usually accomplished by mixtures of nitric, sulfuric and perchloric acids.

Elements receiving widespread attention from the standpoint of nutrition and also health hazards are Hg, Pb, As, Cd, Se, Cr, Co, Ni, Zn, Sb, Cu, Sn and Mn.

DETERMINATION OF METALS IN FISH

This procedure is particularly useful for small whole fish where an analysis of the entire fish is desired.

Procedure

Sample Pretreatment.—Select about 10-15 (or an appropriate number for a composite sample) small fish, wash with deionized water, freeze and lyophilize for approximately 48-72 hr. Grind the lyophilized samples in an all glass system to obtain a powdered homogeneous sample. Place approximately 0.3-0.4 g (weighed analytically) of the sample (in duplicate) in a previously dried and weighed Vycor crucible. Heat in a 105°C oven for 5 hr. Cautiously remove the crucibles from the oven and let cool in a desiccator, then weigh and record the "dry weight" of the sample. Then place the crucibles containing the samples in a muffle furnace. Turn on the temperature control and adjust to 475°C and allow the samples to ash for 16 hr. Remove the samples from the furnace, let cool and add 1.0 ml of concentrated nitric acid to each sample, then allow to stand for 1 hr with frequent stirring during this period. Finally, transfer the samples (analytically) from the crucible to a 25 ml volumetric flask and make up to 25 ml using deionized water. Prepare standards such that they contain 1.0 ml of nitric acid per 25 ml (thus the standards and unknowns have the same amount of nitric acid present). Further dilution may be necessary for elements such as zinc. This method has been used for Cu, Zn, Cd, Ni and Mn.

Analysis by Atomic Absorption.—Follow the manufacturer's recommendation for preparing standards and for the operation of the atomic absorption spectrophotometer. It is suggested that background correction be used if this is an accessory on the particular instrument used. Elements that have resonance lines in the far ultraviolet region of the electromagnetic spectrum show considerable nonspecific background absorption interference which may be mistaken for specific absorption by the element being analyzed. Instruments equipped with background correction (e.g., deuterium) can automatically compensate for the nonspecific absorption. Other techniques for correction can be used if the instrument is not so equipped. Consult the instruction manual of the instrument.

Results.—Results should be reported as μg of element per gram (dry weight) of fish. Also if μg of element per gram (wet weight) is desired then the weight of the original wet sample should be determined prior to drying.

(Note: We have found this method particularly useful in studies of fish in waters contaminated with metals from plating and metal-treating industries. Normal fish would not contain appreciable quantities of cadmium. The instructor may wish to adulterate the fish with a known quantity of cadmium to illustrate the actual contamination.)

DETERMINATION OF METALS IN TEA

Procedure

Ashing.—Weigh approximately 2.0 g of both leaf and instant tea (preferably the same brand) into Vycor crucibles in duplicate. Place in a cold muffle furnace and heat at 500°C until a gray ash is formed (approximately 8-12 hr). Allow the furnace to cool and remove the crucible. Moisten the ash with 10 ml of 8 *M* nitric acid and evaporate on a steam bath to near dryness. Quantitatively transfer the moist residue to a 25 ml volumetric flask and dilute to mark. Make final dilutions based on the concentration of metal being analyzed.

Analyze for Mn, Ca and Na. The specific operation of the atomic absorption spectrophotometer will be found in the operating manual of the particular instrument.

Brewing.—Weigh the contents of a regular size tea bag. Bring 236 ml (1 cup) of deionized water to boil, add the tea and allow the tea to steep for 10 min with stirring. Also weigh the quantity of instant tea (preferably the same brand) recommended for 1 cup and add to 236 ml of boiling water, allow to steep for 10 min. Cool each sample and make the appropriate dilutions as determined in the ashing procedure. Again determine the concentration of Mn, Ca and Na.

Results.—Report the results in each experiment (ashing and brewing) as per cent of the original sample and also μg per gram of original sample. Compare the ashing data to the extraction by brewing. Account for the difference. Also account for the difference in Ca and Na in tea leaves and instant tea.

ANALYSIS OF MERCURY IN FISH USING A FLAMELESS ATOMIC ABSORPTION TECHNIQUE[1]

With the introduction of flameless atomic absorption techniques by Hatch and Ott (1968), considerable attention has been given to elements such as mercury as a contaminant in foods, particularly fish and other marine foods. Instances of mercury poisoning such as that observed in Minamata, Japan, lead to interest in the mercury contamination of foods. The method described is generally useful for 0.001 to 1.0 ppm mercury in a 0.5-1.0 g sample.

Procedure

Reagents

(1) Potassium permanganate crystals and 5%(w/v) solution (Hg free)
(2) 5.6 *N* nitric acid
(3) 18 *N* sulfuric acid
(4) 1.5% hydroxylamine hydrochloride solution
(5) 10% stannous chloride solution
(6) Concentrated sulfuric acid
(7) Mercury standard

Sample Pretreatment.—A 1.0 g sample of homogenized fish tissue is weighed into a 125 ml Erlenmeyer flask. Slowly add 30 ml of concentrated sulfuric acid. Stopper loosely with a polyethylene stopper and allow to stand at room temperature for approximately 15 min. Swirl the contents to disperse and then place in a 50-60°C water bath for at least two hours. If the colored solution contains undissolved matter after the two hour digestion time, add an additional 5 ml of concentrated sulfuric acid and heat for an additional hour.

Cool to room temperature and carefully transfer to a 300 ml BOD bottle containing 50 ml of Hg-free distilled water. Rinse the flask with 20 ml of Hg-free distilled water (two 10 ml rinses) and add the rinses to the BOD bottle. Slowly add potassium permanganate crystals to the bottle. Heat in a 50-60°C water bath. The sample will turn brown and froth. When frothing subsides, add more potassium permanganate until the purple color persists. Swirl the sample throughout the addition.

Analysis[2].—To the BOD bottle containing 100 ml of the prepared sample (Fig. 8.1) add 5 ml of 5.6 *N* nitric acid and mix well. Wait approximately 15 sec, then add 5 ml of 18 *N* sulfuric acid and again mix well. Wait 45 sec. Add 5 ml of hydroxylamine hydrochloride solution and swirl. The sample should turn clear in approximately 15 sec. If not, add hydroxylamine hydrochloride crystals until a clear, colorless solution is obtained. Add 5 ml of stannous chloride solution and turn the air flow on and immediately insert the aerator into the BOD bottle. No appreciable undissolved matter should be present. If matter is noted, a new sample should be prepared.

Record the absorbance reading as the mercury is aspirated into the cell.

Preparation of Standards and Blanks

Prepare a reagent blank that contains all reagents but omits the fish sample and should be treated exactly as the unknown samples.

Standards should be prepared by adding 2 drops of potassium permanganate solution to each of six 300 ml BOD bottles. Prepare duplicate standards containing 0.0, 0.5 and 1.0 μg mercury in a total

[1] Perkin-Elmer Corp., Norwalk, Conn.

[2] These instructions are for the Perkin-Elmer (303-0830) Mercury Analysis System. Other systems may vary; thus their instructions should be consulted.

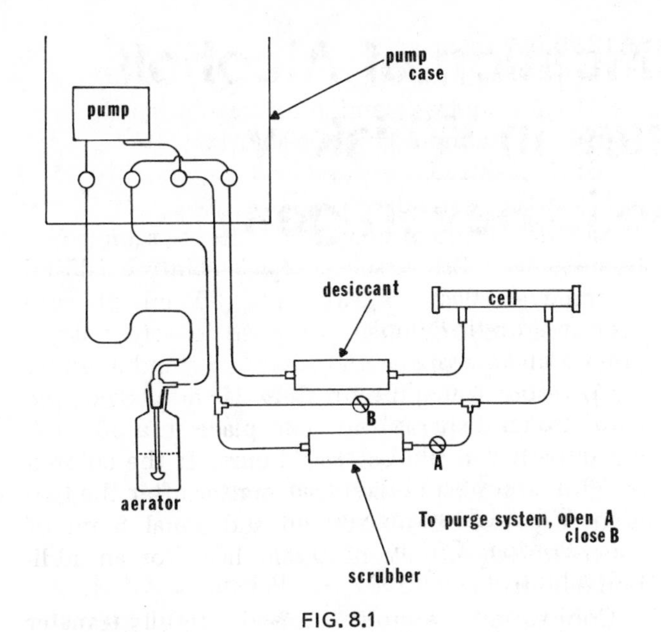

FIG. 8.1

DIAGRAM FOR FLAMELESS MERCURY DETERMINATION

volume of 100 ml. These serve as a blank and two standards. They should be treated exactly as those in the section on *Analysis*.

Results.—Prepare a standard curve of μg of mercury versus maximum absorbence for each mercury standard. Read μg of mercury in the unknown directly from the standard curve and subtract the value obtained for the reagent blank.

Report the results in μg mercury/g of tissue. This would give the concentration of mercury in μg/g (wet weight). If μg/g (dry weight) is desired, then the dry weight can be determined as described in the previous experiment.

METALS IN BEER

Procedure

Sample Pretreatment.—Remove the carbon dioxide by filtering through filter paper or by agitation and warming.

Analysis.—Aspirate the samples directly into the flame. Dilute with deionized water if necessary to bring the concentration of the element of interest into a range suitable for atomic absorption. A three-slot burner is recommended to prevent sample solids from clogging the burner.

This method can be used for calcium, iron, sodium, potassium, copper and other elements. Report the results as μg/ml of original beer.

QUESTIONS

(1) In metal analyses, why is it advisable to determine the percentage recovery of a known quantity of added metal?

(2) What is the function of stannous chloride in the mercury analysis?

(3) Why would metals such as Hg, Cd and Pb be toxic at very low levels?

Separation and Identification of Alcohols and Other Volatiles in Whiskey and Other Alcoholic Beverages

Alcoholic beverages have a variety of volatile components that range in boiling points from 21° C (acetaldehyde) to 219° C (phenyl ethyl alcohol). These components have successfully been separated and identified by gas chromatography; however, a complete separation of the volatiles requires different columns and/or operating conditions.

Often the components in the various alcoholic beverages are divided into classes based on their volatility, e.g.:

(1) Light Ends—Components which elute from the column before ethanol. These generally include acetaldehyde, ethyl acetate, acetone, methanol, and acetol.

(2) Ethanol—The major component in all alcoholic beverages.

(3) Fusel Oils—This fraction includes *n*-propyl, isobutyl, active amyl and isoamyl alcohols.

(4) Components less volatile than isoamyl alcohol—This fraction would include $C_6 - C_{18}$ fatty acid ethyl esters and phenyl ethyl alcohol.

In order for the light ends to be separated and the fusel oils to elute in a reasonable length of time, the gas chromatograph often must be temperature programed. In the following experiments a Poropak Q (Trademark, Waters Associates) column is used to demonstrate the overall procedure. This column packing is a porous polymer of ethylvinylbenzene cross-linked with divinyl benzene. This material allows the separation of many of the volatiles but additionally it has been used quite successfully for the separation and quantitation of water-ethanol mixtures.

Quantitative analysis is widely used in gas chromatography. Internal standardization appears to be most useful in determining the concentration of components in alcoholic beverages. In internal standardization, the exact sample size need not be known. Internal standardization is applicable over a wide range of concentration, and precision can be achieved even when operating conditions may be changed.

Procedure

Reagents and Equipment:

(1) Ethanol
(2) *n*-propyl alcohol
(3) Ethyl acetate
(4) Isobutyl alcohol
(5) Isoamyl alcohol
(6) *n*-butyl alcohol
(7) Whiskey and brandy samples
(8) 5 μl microsyringe (Hamilton)
(9) Poropak Q column (6 ft)

Qualitative analysis.—A Beckman GC-4 gas chromatograph equipped with both thermal conductivity and flame ionization detectors is described. Any gas chromatograph so equipped can be used. Details of operation of the instrument can be obtained from the instruction manual. In this experiment the following operating conditions were used:

Column temperature	220° C
Inlet temperature	240° C
Detector line temperature	250° C
Flow	40 ml/min
Chart speed	1 in/min
Thermal conductivity detector	225 mamp
Gas	Helium

Prepare a synthetic standard mixture consisting of equal moles of water, ethanol, ethyl acetate, *n*-propyl alcohol, isobutyl alcohol, and isoamyl alcohol. Inject this mixture (0.5 μl or appropriate dilution) and attenuate the chromatograph such that all components are on scale. Determine the retention time for each component. (Note: Adjustment of operating parameters may be necessary to achieve separation.) After achieving separation of the known compounds, inject an appropriate quantity of whiskey, bourbon or brandy. In these beverages the water and alcohol content will be much in excess of the concentration of the other components (fusel oils); therefore, attenuation will be necessary in order to detect and measure these minor components. An attenuation of 100X is usually sufficient but may vary with the particular gas chromatograph. Repeat the procedure using the flame ionization detector. Report the results as retention time for each component. Furnish the instructor with a copy of the actual chromatogram properly labeled with all conditions of operation.

Quantitative analysis.—Two separate determinations will be described:

(1) Analysis of the ethanol-water ratio in alcoholic beverages necessitates the use of a thermal conductivity detector since the flame ionization detector does not respond to water. Prepare standard mixtures of ethanol (absolute) and water ranging from 35:65 (v/v) ethanol-water to 65:35. Inject the standards and an equal quantity of an unknown and attenuate the chromatograph so that both the water and alcohol peaks are on scale. Determine the ratio of the alcohol to water from the standard mixtures.

(2) Fusel oils include *n*-propyl, isobutyl, isoamyl and active amyl alcohol. These may conveniently be detected using a flame ionization detector. Prepare three standard mixtures in 40% ethanol in water as follows:

Standard	*n*-propyl	ppm isopropyl	isoamyl
1	100	100	200
2	300	300	600
3	600	600	900

Add *n*-butyl alcohol (200 ppm final concentration) to each mixture to act as an internal standard. Chromatograph the standard mixtures (adjustment of operating parameters may be necessary) such that all peaks are sharp and distinctly resolved. Plot a calibration curve of the concentration of each component (Y axis) versus the solute peak height/internal standard peak height (X axis). Add enough *n*-butyl alcohol to the unknown to give a final concentration of 200 ppm and chromatograph the unknown. Calculate the ratio of the individual fusel oils in the unknown to the peak height of the internal standard and compare to the standard graph to determine the concentration of these components in the unknown. (Note: If alcoholic beverages contain concentrations higher than the standard mixtures, the upper range of standards should be increased.)

QUESTIONS

(1) Why must different standard curves be constructed for each of the fusel oil alcohols?

(2) What advantages does Poropak Q offer for alcohol-water separations?

(3) Suggest a set of conditions and a column that would allow all of the volatiles in bourbon to be separated.

Calorimetry

There are a number of chemical changes that occur during the absorption and utilization of food. These chemical changes are coupled with energy transformations, most commonly noted by the production of heat. There are, however, other types of energy transformations involved in metabolism. Chemical changes that are oxidative in nature and are characterized by development of heat are referred to as exothermic reactions. Hydrolysis reactions are characterized by little or no heat evolution and as a consequence are referred to as isothermic reactions. If heat is absorbed during the chemical change, the reaction is referred to as an endothermic reaction. Reduction reactions are endothermic reactions.

Animals, in contrast to plants, must depend upon presynthesized foods; therefore exothermic reactions predominate in the animal body during metabolism. Carbohydrates, fat and proteins are the nutrients in food which are essential in energy production. It has been found that calorimetry is very useful in making an assessment of the potential nutritive value of foods. To do this, it is necessary to use some unit by which the potential energy of food can be measured. The unit commonly used in expressing energy value of foods is the large Calorie (C) which is the equivalent to 1000 calories. A small calorie is defined as the amount of heat required to raise the temperature of 1 g of water 1°C.

Procedure

The following description is essentially taken from Manual No. 147 furnished by Parr Instrument Company for their 1341 plain jacket oxygen bomb calorimeter as shown in Fig. 10.1. (Consult instruction manuals for other calorimeters.) The operating principle for this calorimeter is the same as in all bomb calorimeters. A weighed sample is burned in an oxygen-filled metal bomb while the bomb is held in a measured quantity of water within the thermal insulating jacket. By observing the temperature rise of the water and knowing the energy equivalent of the calorimeter, the amount of heat released from the sample can be calculated. Test results are commonly expressed in calories per gram (cal/g), British thermal units per pound (Btu/lb), or in the large Calorie per gram units commonly used for foods. They may also be expressed in joules per kilogram (J/kg) as used in the SI system of international units. In this manual the International Table calorie is used. This is defined as equal to 4.1868 joules. The following conversion factors may be used:

1 cal/g = 1.8 Btu/lb
1 cal/g = 0.001 Cal/g
1 cal/g = 4186.8 J/kg

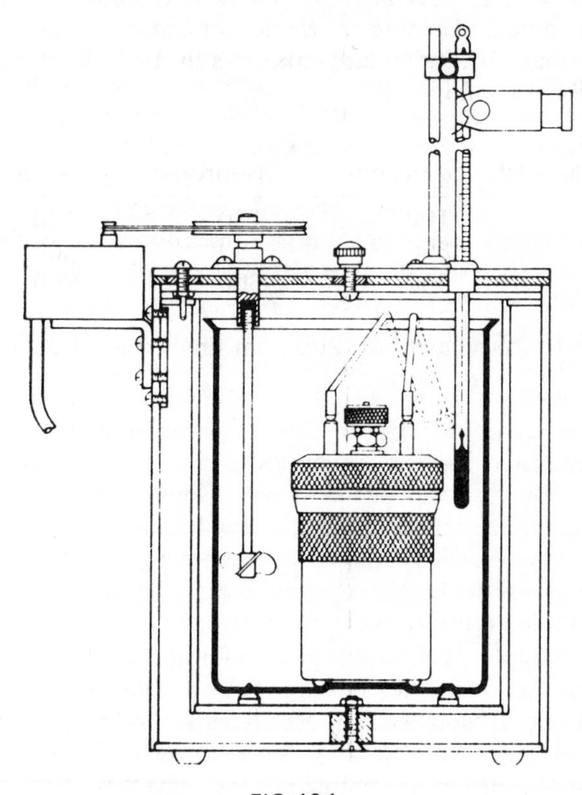

FIG. 10.1
CROSS-SECTION OF PARR 1341 CALORIMETER

Samples And Sample Holders

Selection and Preparation of Solid Samples.— Solid samples should be air-dried and ground until all particles will pass through a 60-mesh screen. The particle size is important because the combustion reaction proceeds to completion within a few seconds and if any of the individual particles are too large they will not burn completely. A sample that is too finely divided may also be difficult to burn because extremely small particles can be swept out of the combustion capsule by the turbulent gases. If they fall to the bottom of the bomb without being ignited, the test will be unsatisfactory. The Parr pellet press offers a possible solution to the problem of incomplete combustion when working with powdered or finely divided samples. Some operators prefer to work with moisture-free or bone-dry samples instead of using air-dried samples as suggested above. There is no objection to this procedure if care is taken to prevent overheating and loss of volatile constituents when preparing the dry sample. Care must also be taken to avoid moisture absorption from the atmosphere when the sample is weighed.

Foodstuffs, and Cellulosic Materials.—The high moisture content of foodstuffs may require that they be dried before making a calorific test, but this is not true in all cases. Samples containing up to 40% moisture can usually be burned in an oxygen bomb without preliminary drying, but samples with higher moisture contents may have to be dried. If possible, use a portion of the sample as received for the calorimetric test and run the moisture determination on a separate portion. When dealing with new materials it may be necessary to make several preliminary tests to determine the approximate maximum allowable moisture content at which the sample will ignite and burn completely. A certain amount of moisture is desirable in calorimetric samples since a very dry sample may burn so rapidly that the particles will be carried out of the capsule to strike the cold wall of the bomb, leaving a smudge of unburned carbon as an indication that complete combustion was not obtained. Rapid burning rates can be slowed by adding a small amount of water to the weighed sample and allowing time for the moisture to be absorbed uniformly throughout the sample before firing the bomb.

Sample Pellets.—It is helpful to compress powdered samples into a pellet or tablet before the sample is weighed. Pellets are easier to handle than loose samples and they burn slower in an oxygen bomb, thereby reducing the chances for incomplete combustion. They are usually made in a Parr pellet press with one-half inch diameter punch and die. Instructions are supplied with each press. It will be convenient to make several pellets from the same sample and hold them in a stoppered vial or weighing bottle until they are to be weighed into a fuel capsule. Pellets should be handled with forceps or a small pair of tongs and not touched with the fingers.

(Note: A pellet of sucrose or other sugar is a convenient sample to demonstrate the use of oxygen bomb calorimetry.)

Charging the Bomb

Allowable Sample Size.—Care must be taken to avoid overcharging the bomb for it must be realized that the peak pressure developed during a combustion is proportional to the size of the sample and to the initial oxygen pressure. To stay within safe limits, the bomb should never be charged with a sample which will release more than 8000 calories when burned in oxygen, and the initial oxygen pressure should never exceed 35 atmospheres (515 psig). This generally limits the mass of the combustible charge (sample plus benzoic acid, gelatin, firing oil or any combustion aid) to not more than 1.1 grams. When starting tests with new or unfamiliar materials it is always best to use samples of less than one gram, with the possibility of increasing the amount if preliminary tests indicate no abnormal behavior. To avoid damage to the bomb and possible injury to the operator, the bomb must never be charged with more than 1.5 g of combustible material.

Attaching the Fuse.—Set the bomb head on the A38A support stand and attach a 10 cm length of fuse wire as shown in Figure 10.2. Parr 45C10 nickel alloy wire will be used for most tests, but the same procedure applies when using 36 gauge or finer platinum wire. In either case, a pair of forceps or tweezers will be helpful for binding the wire to the electrodes. First, fasten one end of the wire to the loop electrode (steps e through h). Pull the loop downward to tighten the connections; then bend the wire upward as in detail i. Place the capsule in the loop holder and bend the wire down to touch the surface of the charge as shown in detail j.

It is not necessary to submerge the wire in a powdered sample. In fact better combustions will usually be obtained if the loop of the fuse is set slightly above the surface. When using pelleted samples, bend the wire so that the loop bears against the top of the pellet firmly enough to keep it from sliding against the side of the capsule. It is also good practice to tilt the capsule slightly to one side so that the flame emerging from it will not impinge directly on the tip of the straight electrode.

Water in the Bomb.—Place 1.0 ml of distilled water in the bomb from a pipet.

Closing the Bomb.—Care must be taken not to disturb the sample when moving the bomb head from the support stand to the bomb cylinder. Be sure that the 104A2 contact ring is in place above the sealing ring and that the sealing ring is in good condition; then slide the head into the cylinder and push it down as far as it will go. It will be convenient (but not essential) to hold the bomb in an A124A2 bench clamp during the closing operation and while filling the bomb with oxygen. Set the screw cap on the cylinder and turn it down firmly by hand. Do not use a wrench or spanner on the cap. Hand tightening should be sufficient to secure a tight seal.

Oxygen Filling Equipment.—Commercial oxygen as supplied in a standard 1A cylinder with CGA No. 540 outlet is usually of sufficient purity for calorimetric work. Connections to the oxygen cylinder are made with a Parr 1825 filling connection which is supplied with the calorimeter. This connection has a needle valve which controls the flow into the bomb and a 0-55 atm gauge which shows the pressure to which the bomb has been charged. It also has an automatic relief valve to prevent overcharging. To attach the filling connection, unscrew the protecting cap from the oxygen tank and inspect the threads on the valve outlet to be sure they are clean and in good condition. Place the ball end of the connection into the cylinder outlet and draw up the union nut tightly with a wrench, keeping the 0-55 atm gauge in an upright position.

Filling the Bomb.—Press the fitting on the end of the oxygen hose into the inlet valve socket and turn the knurled union nut finger tight. Close the control valve on the filling connection; then open or "crack" the oxygen tank valve not more than one-quarter turn. Open the filling connection control valve slowly and watch the gauge as the bomb pressure rises to the desired filling pressure (usually not more than 30 atm); then close the control valve. The bomb inlet check valve will close automatically when the oxygen supply is shut off, leaving the bomb filled to the highest pressure indicated on the 0-55 atm gauge. Release the residual pressure in the connecting hose by pushing downward on the lever attached to the relief valve. The gauge should now return to zero. If the pressure drops slowly and a large amount of gas escapes when the release valve is opened, the check valve in the bomb head is not operating properly. This trouble will have to be corrected before the bomb can be used.

Safety Requirements

The high pressures and nearly explosive reactions which occur within an oxygen bomb need not be considered unusually hazardous, provided that the bomb is in good operating condition and that the operator follows the recommended test procedure. All persons concerned with safe operation of the calorimeter should insist upon compliance with the directions given herein. Particular emphasis should be given to the following basic safety requirements:

(1) Do not use too much sample. The total

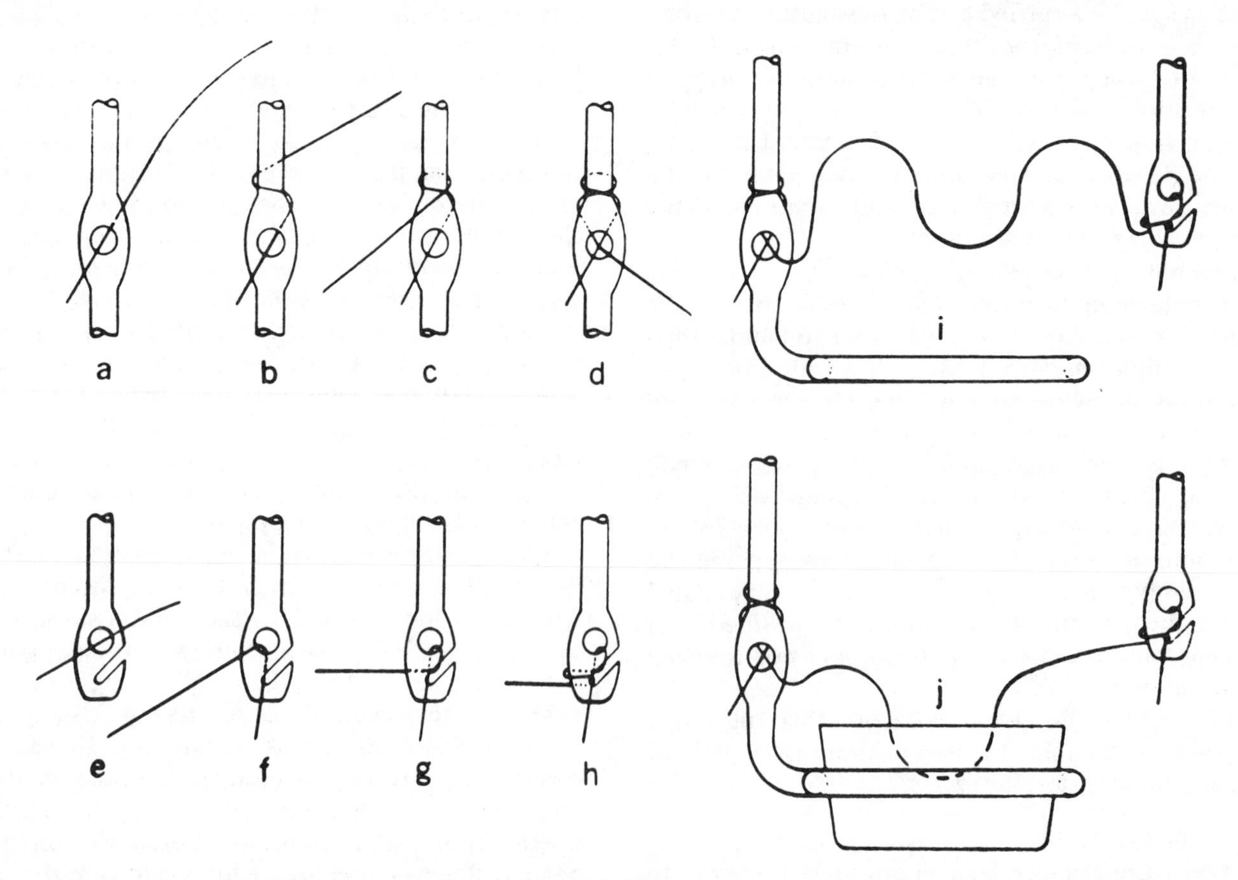

FIG. 10.2

STEPS IN BINDING FUSE WIRE TO 4A and 5A ELECTRODES

charge placed in the Parr 1108 bomb must not liberate more than 8000 calories when burned in oxygen at an initial pressure of 35 atmospheres. This usually limits the sample weight to not more than 1.1 g.

(2) Use a small fraction of the maximum allowable sample weight when testing unfamiliar materials which may burn either rapidly or explosively.

(3) Do not fill the bomb with more oxygen than is necessary to obtain complete combustion. Do not fire the bomb if it is pressurized to more than 35 atmospheres.

(4) Keep all parts of the bomb, especially the valves and insulated electrode, in good repair at all times. If gas bubbles escape from the bomb when it is submerged in water, do not fire the charge.

(5) Keep away from the top of the calorimeter for 30 sec after firing. If the bomb should rupture, it is most likely that the resultant forces will be directed along the vertical axis.

(6) Never use lubricants on valves or fittings in contact with high pressure oxygen.

Operating The Calorimeter

Filling the Bucket.—Weigh the dry calorimeter bucket on a solution or trip balance, then add 2000 (plus or minus 0.5) grams of water. Distilled water is preferred, but demineralized or tap water containing less then 250 ppm of dissolved solids is satisfactory. The water temperature should be approximately 1.5° C below room temperature, but this can be varied to suit the operator's preference. The bucket may be filled from an automatic pipet or from any other volumetric device if the water is measured with comparable precision.

Assembling the Calorimeter.—Set the filled bucket in the calorimeter with the bomb-locating boss in the bottom of the bucket facing the front of the calorimeter. Attach the lifting handle to the two holes in the side of the screw cap and lower the bomb into the water. Before the head is submerged, push the ignition wires into the two terminal sockets on the bomb head; then push any excess wire back through the hole in the jacket liner. Lower the bomb into the water with its feet spanning the boss in the bottom of the bucket. Handle the bomb carefully during this process so that the sample will not be disturbed. Remove the lifting handle and shake any drops of water back into the bucket; then set the cover on the calorimeter with the thermometer facing toward the front and with the cover-locating pin inserted into the hole in the jacket rim. Turn the stirrer by hand to be sure that it runs freely; then attach the drive belt and start the motor. The calorimeter is now ready for the test run.

The Test Run.—Let the stirrer run for 5 min to reach thermal equilibrium before starting a measured run. At the end of this period, record the time or start a timer and read the temperature to the nearest 0.002° C (or to 0.005° F if a Fahrenheit thermometer is used). Continue to read and record the temperature at one-minute intervals for exactly 5 min; then press the bottom on the ignition unit to fire the charge at the start of the sixth minute. Always tap the thermometer with a pencil or rod to vibrate the mercury before taking a reading, or use a Parr 3010 thermometer vibrator for this purpose. The vibrator sits on top of the thermometer stem and exerts a series of sharp impulses to level the mercury meniscus when activated by a control button on an A84C power supply.

At approximately 20 sec after ignition the temperature will begin to rise. The rate of rise will be large at first and then decrease as the bomb, water and bucket approach a new equilibrium temperature. The nature of the temperature rise is indicated by the typical temperature rise curve shown in Figure 10.3. It is not necessary to plot a similar curve for each test, but accurate time and temperature observations must be recorded to identify the critical points needed to calculate the calorific value of the sample or the energy equivalent for the calorimeter.

If the net temperature rise can be estimated from previous tests with similar samples, add 60% of the expected total rise to the observed temperature at ignition and locate this point on the thermometer scale. Record the time in minutes and decimal fractions of a minute when the mercury column reaches this temperature. This observation can be taken without the thermometer magnifier since the temperature will be rising rapidly at this point and it will be difficult to keep the magnifier in focus.

If the net temperature rise at the 60% point cannot be estimated before ignition, the time required to reach the 60% point can be found by linear interpolation from readings taken during the rise period. This requires temperature observations at 45, 60, 75, 90 and 105 sec after firing. These can be taken without a magnifier since readings estimated to the nearest 0.02° are sufficient at this point.

After the rapid rise period (about 4 or 5 min after ignition) adjust the reading lens and record temperatures to the nearest 0.002° C (0.005° F) at one-minute intervals until the difference between successive readings has been constant for five min. Usually the temperature will reach a maximum; then drop very slowly. But this is not always true since a low starting temperature may result in a slow continuous upward rise without reaching a maximum. As stated above, the difference between successive readings must be noted and the readings continue at one-minute intervals until the rate of the temperature change becomes constant over a period of 5 min.

Opening the Calorimeter.—After the last temperature observation has been made, stop the motor, remove the belt and lift the cover from the calorimeter. Wipe the thermometer bulb and the stirrer with a clean cloth to remove any drops of water and set the cover on the A37A support stand. Lift the bomb out of the bucket; remove the ignition leads and wipe the bomb with a clean towel.

Open the knurled valve knob on the bomb head to release the residual gas pressure before attempting to remove the cap. This release should proceed slowly over a period of not less than one minute to avoid entrainment losses. After all pressure has been released, unscrew the cap; lift the head out of the cylinder and place it on the support stand. Examine the interior of the bomb for soot or other evidence of incomplete combustion. If such evidence is found, the test will have to be discarded.

Note: At this point in the procedure the calculations may be simplified if the original sample has negligible quantities of nitrogen and sulfur. The fuse correction also may be omitted if the fuse wire is 36 gauge or finer platinum. Additionally a simplified procedure can be used for determining T_o (temperature at time of ignition) and T_i (temperature at end of combustion).

The acid titration for HNO_3 and H_2SO_4 may be omitted if the sample has negligible or very low quantities of nitrogen and sulfur.

After constructing a temperature rise curve as shown in Fig. 10.3, extrapolate line c,d back toward the Y axis to intersect with a vertical line (parallel with Y axis) beginning at point a (which corresponds with the time of ignition).

The simplified equation for calculating the heat of combustion is then:

$$\Delta E = \frac{C_{(s)}\,(T_o - T_i)}{\text{wt of sample (g)}}$$

where

ΔE = Heat of combustion; cal/g (negative sign indicates an exothermic reaction)

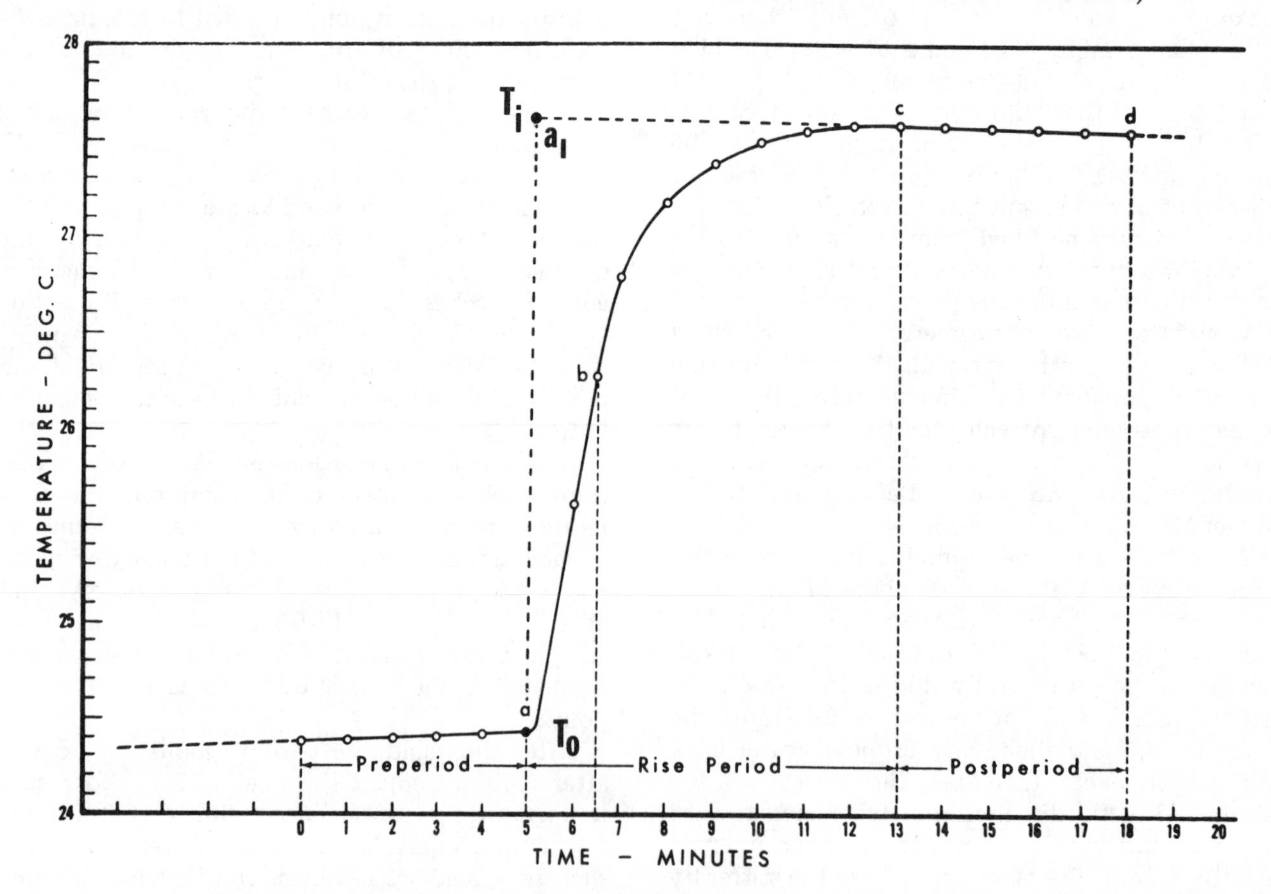

FIG. 10.3
TYPICAL TEMPERATURE RISE CURVE FOR 1341 PLAIN CALORIMETER

$C_{(s)}$ = Heat capacity = calories necessary to change the water temperature 1°C. This is a constant for a specific amount of water and the calorimeter, *per se.*

T_o = Initial temperature of water bath at point a in Figure 10.3

T_i = Final temperature of water bath at point a_1 in Figure 10.3

A more precise heat of combustion, H_g, can be determined by following the procedure listed below.

The Acid Titration.—Wash all interior surfaces of the bomb with a jet of distilled water and collect the washings in a beaker. Titrate the washings with a standard sodium carbonate solution using methyl orange or methyl red indicator. A 0.0725 *N* sodium carbonate solution is recommended for this titration to simplify the calculation. This is prepared by dissolving 3.84 grams of Na_2CO_3 in water and diluting to one liter. NaOH or KOH solutions of the same normality may be used.

The Fuse Correction.—Carefully remove all unburned pieces of fuse wire from the bomb electrodes, straighten them and measure their combined length in centimeters. Subtract this length from the initial length of 10 centimeters and enter this quantity on the data sheet as the net amount of wire burned. (This correction may be omitted when using 36 gauge or finer platinum fuse wire.)

The Sulfur Correction.—The sulfur content of the sample should be determined if it exceeds 0.1%. This analysis can be made by any standard gravimetric, volumetric or nephelometric method for sulfur using the solution remaining from the acid titration. A suggested procedure is given on page 15 of the Parr Manual #147, also in ASTM Method D271, Sections 22 and 23.

Calculating The Heat Of Combustion

Assembly of Data.—The following data should be available at the completion of a test in a 1341 calorimeter:

a = time of firing

b = time (to nearest 0.1 min) when the temperature reaches 60% of the total rise

c = time at beginning of period (after the temperature rise) in which the rate of temperature change has become constant

t_a = temperature at time of firing, corrected for thermometer scale error

t_c = temperature at time c, corrected for thermometer scale error

r_1 = rate (temperature units per minute) at which temperature was rising during the 5 min period before firing

r_2 = rate (temperature units per minute) at which the temperature was rising during the 5 min period after time c. If the temperature was falling instead of rising after time c, r_2 is negative and the quantity $-r_2(c-b)$ becomes positive and must be added when computing the corrected temperature rise.

c_1 = milliliters of standard alkali solution used in the acid titration

c_2 = percentage of sulfur in the sample

c_3 = centimeters of fuse wire consumed in firing

W = energy equivalent of the calorimeter, determined under standardization

m = mass of sample in grams

Temperature Rise.—Compute the net corrected temperature rise, t, by substituting in the following equation:

$$t = t_c - t_a - r_1(b-a) - r_2(c-b)$$

Thermochemical Corrections.—Compute the following for each test:

e_1 = correction in calories for heat of formation of nitric acid (HNO_3)
= c_1 if .0725 *N* alkali was used for the titration

e_2 = correction in calories for heat of formation of sulfuric acid (H_2SO_4)
= (14) (c_2) (m)

e_3 = correction in calories for heat of combustion of fuse wire
= (2.3) (c_3) when using Parr 45C10 nickel-chromium fuse wire, or
= (2.7) (c_3) when using No. 34 B. & S. gauge iron fuse wire

Gross Heat of Combustion.—Compute the gross heat of combustion, H_g, in calories per gram by substituting in the following equation:

$$H_g = \frac{tW - e_1 - e_2 - e_3}{m}$$

Example:

a = 1:44:00 = 1:44.0
b = 1:45:24 = 1:45.4
c = 1:52:00 = 1:52.0
t_a = 24.428 + .004 - 24.432°C
t_c = 27.654 + .008 = 27.662°C
r_1 = +0.010°C/5 min = +0.002°C/min
r_2 = -0.004°C/5 min = -0.001°C/min
c_1 = 23.9 ml
c_2 = 1.02% Sulfur
c_3 = 7.6 cm Parr 45C10 wire
W = 2426 calories/°C
m = 0.9936 grams

t = 27.662 - 24.432 - (0.002) (1.4) - (-0.001) (6.6)
= 3.234°C

e_1 = 23.9 calories

e_2 = (14) (1.02) (0.9936) = 14.2 calories

e_3 = (2.3) (7.6) = 17.5 calories

$$H_g = \frac{(3.234)(2426) - 23.9 - 14.2 - 17.5}{0.9936}$$

= 7840 calories/gram

= (1.8) (7840) = 14,110 Btu/lb

= (4186.8) (7840) = 3.282×10^7 joules/kg

Anthocyanins

Anthocyanins are the major color-imparting compounds in foodstuffs such as cherries, cranberries, grapes, strawberries, currants, raspberries, boysenberries, eggplant, just to name a few. Anthocyannis (Gr. *antho*, flower, and Gr. *kyanos*, blue) consist of two major structural features, the sugar moiety and an aglycon (anthocyanidin) which can be separated by acid hydrolysis. A variety of sugars are found in the glycosides. These include mono-, di-, and trisaccharides. The sugars are generally attached to the anthocyanidin, flavone, flavonol or flavanones by a β-glycosidic linkage. D-glucose, D-galactose, L-rhamnose, D-arabinose and D-xylose are the most commonly occurring monosaccharides. Disaccharides include rutinose or L-rhamnosyl ($\alpha 1 \rightarrow 6$) D-glucose, gentiobiose or D-glucosyl ($\beta 1 \rightarrow 6$) D-glucose, sophorase or D-glucosyl ($\beta 1 \rightarrow 2$) D-glucose, and sambubiose or D-xylosyl ($\beta 1 \rightarrow 2$) D-glucose. The carbohydrate moieties are attached most commonly to the 3-hydroxyl position of the anthocyanidin. The sugars in diglycosides are attached to either the 3 and 5 hydroxyl groups or the 3 and 7 hydroxyl groups.

ANALYTICAL TECHNIQUES FOR ANTHOCYANINS AND FLAVONOIDS

The most used technique for removing the colored anthocyanins and flavonoids from their natural source has been water extraction. The AOAC method (1970, 22.102) utilizes such a procedure for removing the anthocyanins from grape and various colored fruit juices. Various other techniques have utilized methanol, ethanol, and other alcohols for removing the pigments.

Paper chromatography has been the most popular method for separating and identifying anthocyanins and flavonoids. Commonly, the extracted pigments are applied directly to chromatography paper (Whatman No. 1 and others) followed by development in one or more solvent systems. The most used solvent mixture is butanol, acetic acid, and water of varying ratios.

Final identification is achieved by a number of procedures. Comparison of R_f values of the unknown or suspected anthocyanins, anthocyanidins, or flavonoids, to those of known compounds, is the method of choice for chromatographic and electrophoretic techniques.

Further identification of the bands, spots, or zones on the separating media can be achieved by eluting the pigment from the media, dissolving in an appropriate solvent, and determining the absorption spectra. Table 11.1 gives the wavelength of maximum absorption for some of the flavonoids. The data in this table give only the range of absorption maxima.

The three major anthocyanidins (aglucons) and their glycosides which appear in almost all foodstuffs are usually identified specifically by determining the wavelength of maximum absorption in the visible region of the spectra. The table gives these maxima and the color imparted by the compounds. Although numerous other techniques for positive characterization are available, the data in the table represent the most commonly used ones.

TABLE 11.1
WAVELENGTH MAXIMA FOR MAJOR ANTHOCYANIDINS AND ANTHOCYANINS

Compound	max[1] (nm) (Aglucon)	max[2] (nm) (3-Glycoside)	Color Imparted
Cyanindin	535	507	Magenta
Delphinidin	545	516	Mauve
Pelargonidin	520	492	Orange-red

[1] Determined in methanol-HCl
[2] Determined in aqueous HCl

Procedures

Extraction of Anthocyanins in Grape Juice and Grape Drink (AOAC Method).—The anthocyanins of Welch's Concord grape juice and Welch's grape drink are compared. Add 10 ml of 8% (w/v) lead acetate to 10 ml of juice or drink in a 50 ml polyethylene centrifuge tube (screw-capped or snap-on cap) and thoroughly mix. Then add 0.5 ml of concentrated ammonium hydroxide (NH_4OH) and mix. Centrifuge until the precipitate is well packed. Discard the clear supernatant (if supernatant is not clear, add more lead acetate solution and recentrifuge). Wash the precipitate two times with 25 ml portions of 80% ethanol. Mix well each time before centrifuging. Discard the washings. After the second wash, invert the tube for 5 min to drain the remaining liquid. Add 10 ml of 1-butanol and 1 ml of concentrated HCl. Shake vigorously until all color is released into the alcohol layer. Centrifuge and decant the clear liquid into a 125 ml separatory funnel. Wash the precipitate with 5 ml of 1-butanol, mix well and centrifuge to obtain a clear upper layer. Add the washings to the first extraction in the separatory funnel. Add 100 ml of petroleum ether to the separatory funnel, shake well, and allow to stand. The anthocyanin concentrates in the colored aqueous layer (the lower layer will be approximately 2 ml). If no aqueous layer separates, add 0.5 ml H_2O, shake and swirl. (Note: This step in the procedure is critical to actually comparing different juices or juices to fruit drinks. Water added throughout the procedure will appear at this point, serving to dilute the concentration of anthocyanins. We have found it feasible to reduce the volume of the final H_2O extracts by rotary evaporation at reduced pressure.) Let the liquid layers separate completely and drain the lower aqueous (colored) layer into a 15 ml conical graduated centrifuge tube. Add 0.2-0.5 ml (see note above) portions of H_2O to the separatory funnel, shaking each time and draining lower (colored) layer, until a total volume of 2.5 ml is collected. (Occasionally, 2.5 ml or more will be collected in the first separation of liquids; make no additional H_2O extractions in this case. In fact you may want to reduce the volume as listed above in the note.) Use 0.5 ml for chromatography of anthocyanins. Retain anthocyanin solution in refrigerator until used.

Preparation of Anthocyanidins.—Add an equal volume of 2 *N* HCl to the colored anthocyanin solution (prepared in previous section) in a small test tube containing a small boiling chip. Fit the tube with an air condenser made from glass tubing and a cork. Place the tube in a boiling water bath and heat for 30 min. Remove from the bath and cool in cold H_2O. Add 1 ml of isoamyl alcohol, shake vigorously, and centrifuge. Transfer the clear, upper layer to a small bottle. (Do not remove any turbid aqueous layer in the tube.) Test the aqueous layer for sugars (see section on carbohydrates). Keep anthocyanidin solution in refrigerator.

Chromatography of Anthocyanins.—Use 46 cm × 57 cm Whatman No. 1 paper and 30 cm diameter × 60 cm height chromatography jars. Chromatograms require overnight development; therefore, preparation in afternoon will permit 24 hr development. This method is definitive for fruit juices other than grape guice. It is not definitive for grape juice where the anthocyanin pattern is extremely complicated. However, it is a useful guide particularly when comparing grape juice to grape drink.

Prepare a streak of the anthocyanin extract one inch above the bottom of the paper and approximately four inches apart. Make the streaks 1.25″ long and 1/8-1/4 inches wide, using a capillary melting point tube drawn to about 1/2 original diameter at one end. Concord grape juice requires 5-7 applications to obtain satisfactory intensity. Grape drink requires many more applications due to the very low concentration of anthocyanins present. Other juices usually require more applications; however, some may require fewer. Make at least two separate streaking from the same solution, using different intensities, one of which may yield better chromatographic pattern. Dry between each application with cold air blast only. After streakings are completed, dry streaks, and make a cylinder of the paper with the streaks at one end of the cylinder. Sew the edges of the cylinder together using a large needle and colorless cotton thread. Leave about 0.5 inches between the edges. Place approximately 0.5 inches of developing solvent in the tank and insert cylinder in such a manner as not to allow the paper to come in contact with the wall of the tank. Let develop until the solvent front has risen to approximately 1-2 inches from the top of the paper. Remove and air dry. Simultaneously test authentic fruit juice anthocyanins and compare patterns. Note differences in natural light and under long wavelength uv light. Spray with phosphomolybdic acid solution and observe changes in the pattern. Foreign natural coloring material is indicated by significant differences from the authentic pattern.

QUESTIONS

(1) Compare the intensity of the color of the

anthocyanin extract from equal volumes of grape juice and grape drink. (a) How do the chromatograms compare in the two extracts? (b) Does the grape drink contain grape juice? (c) How might one prove this?

(2) What was the function of the concentrated ammonium hydroxide in the lead acetate precipitation? Why were both 1-butanol and HCl added to the colored precipitate?

(3) What was the function of the HCl in preparing the anthocyanidin?

(4) Does pH affect the color of anthocyanins (anthocyanidins)? What significance could this have in food products?

Isolation and Purification of Lycopene

Lycopene is a polyunsaturated compound found in tomatoes, fruits and berries. It has a distinctive red color. The formula for lycopene is:

$$[(CH_3)_2C=CH-CH_2-CH_2(-\overset{CH_3}{\overset{|}{C}}=CH-CH=CH)_2-\overset{CH_3}{\overset{|}{C}}=CH-CH=]_2$$

In the isolation, tomato paste is dehydrated with methanol, and lycopene is extracted from the residue with methanol and carbon tetrachloride. The product is recrystallized twice from benzene to give 98-99% purity.

Procedure

Reagents and Supplies

(1) Canned tomato paste (not purée or conserve)
(2) Methanol and carbon tetrachloride
(3) 20 cm Büchner Funnel
(4) 4 liter separatory funnel
(5) 50 ml centrifuge tube

Extraction of Lycopene.—Add 1.3 liter of methanol to 1 kg of tomato paste in a 5 liter container. The mixture is shaken vigorously immediately; otherwise hard lumps will form. If the mixture has a glue-like consistency, add more methanol. This will prevent clogging of filters. Allow the mixture to stand for 1-2 hr followed by vigorous shaking. Filter in a large Büchner funnel (20 cm) using fast flowing filter paper (no filter aid) or centrifuge in a large centrifuge. Discard the yellow filtrate (or supernatant if centrifuged).

The red residue is returned to the original container (5 liter), mixed with 650 ml of methanol and 650 ml of carbon tetrachloride and shaken to extract the lycopene. Care should be exercised since pressure builds up inside the stoppered container and must be carefully released. The suspension is shaken mechanically for 20 min. The mixture is either filtered or centrifuged as before. The filtrate consists of a lower dark red-colored, carbon tetrachloride layer and an orange-colored water methanol layer. The residue is crushed and re-extracted as before with methanol-CCl_4.

The extracts are combined and the methanol layer (upper layer) is transferred to a 4 liter separatory funnel. Add an equal volume of water. A white emulsion forms in the upper layer. Stir the upper layer with a glass rod to break loose any CCl_4 droplets trapped in the upper layer. Siphon off the upper layer to leave the lower red CCl_4 layer containing the lycopene along with some methanol and other impurities which are removed by introducing a glass tube (5 mm diameter) through the top of the separatory funnel and letting a slow stream of water flow through the CCl_4 extract for 10 min. (The funnel can be hung over a sink and the water allowed to overflow along its outside walls.) The CCl_4 extract is transferred to a 2 liter flask and dried over anhydrous sodium sulfate. The extract is filtered into a 2 liter round bottom standard taper flask. The solvent is then evaporated to about 100 ml on a rotary evaporator, under vacuum, at approximately 60° C. The extract is then transferred to a smaller 200 ml flask and the solvent removed completely removed under vacuum as before.

Purification.—The oily residue is diluted with a few ml of benzene and evaporated in order to eliminate the CCl_4. The dark residue is transferred into a 125 ml flask using 25 ml of benzene. Warm the solution in a hot water bath until it becomes clear. Care should be taken to protect the solution from light and it also should be stored in the cold (approximately 4° C) if the procedure should be interrupted. Boiling methanol (15 ml) is added dropwise with stirring. Impure crystals of lycopene should appear immediately. Crystallization is promoted by cooling the solution in an ice bath for 1 to 2 hr. Filter through a small fritted-glass funnel with suction. Wash the crystals five times with 10 ml portions of boiling methanol being careful not to allow the methanol to cool.

Transfer the lycopene crystals to a 50 ml centrifuge tube and again recrystallize as before from 25 ml of benzene by adding methanol. No more than 20 ml of methanol should be added. Cool and allow to stand in an ice bath for 2 hr. Decant the supernatant and discard. The crystals are again treated with 25 ml of boiling methanol and the crystals separated by centrifugation before the methanol cools. Repeat the washing twice more. Collect the crystals, dry under vacuum (0.5-1.0 mm) and weigh.

Product Examination.—Red lycopene prisms should be observed under the microscope. Determine the melting point (literature m.p. 173° C). Determine the visible absorption spectra in petroleum ether. Also determine the infrared absorption spectra and compare to the infrared spectra of authentic lycopene.

QUESTIONS

(1) Where in the purification procedure might the carotenes be removed?

(2) Should special precautions be observed when storing lycopene?

Isolation and Purification of Phosvitin: A Phosphoprotein from Egg Yolk

Phosphate-containing proteins are widespread in foods, e.g., the protein, casein and enzymes such as phosphoglucomutase and phosphoglyceromutase. Phosvitin is a phosphoprotein from egg yolk that can be precipitated with ammonium sulfate.

Procedure

Reagents and Supplies

(1) Fresh egg yolks (250 g from approximately 15 average size eggs)
(2) 1.2 *M* Magnesium sulfate
(3) 0.4 *M* Ammonium sulfate
(4) 6 *N* Sulfuric acid
(5) 1.0 *M* Acetate buffer pH 4
(6) Filter aid
(7) Dialysis tubing
(8) 0.5 *M* Sodium chloride
(9) 2 *M* Sodium chloride
(10) Toluene and ethyl ether

Sample Pretreatment.—The egg yolks are separated, washed in water and rolled on cheesecloth to remove remaining whites. Remove the stringy membranes attached to the yolk. Puncture the yolk membranes, drain and filter through a single layer of cheesecloth.

Isolation of Phosvitin.—Mix 125 ml of 1.2 *M* magnesium sulfate with 250 g of egg yolks. Stir vigorously for 1 hr (do not allow appreciable foaming). Slowly add 1.25 liter of water (containing 75 mg of Merthiolate) with stirring over a period of approximately 1 hr. The mixture is covered and allowed to stand at room temperature overnight (the surface should be covered with toluene). Centrifuge the soft, sticky precipitate and discard the supernatant.

The precipitate is dispersed in 175 ml of 0.4 *M* ammonium sulfate by stirring for approximately 45 min. Adjust the solution to pH 4 with 6 *N* sulfuric acid. With vigorous stirring add 7.5 ml of 1.0 *M* acetate buffer (pH 4) and 150 ml of ether. The mixture is then shaken vigorously. Centrifuge and discard the ether layer. A gelatinous emulsion layer will be observed over the aqueous layer. The aqueous layer, which contains the phosvitin, is separated from the gelatinous emulsion layer. The gelatinous fraction is extracted three more times as follows: (Extraction 1) 65 ml of ether, 100 ml of 0.4 ammonium sulfate, and 4.25 ml of 1.0 *M* acetate buffer; (Extraction 2 and 3) 65 ml of ether, 88 ml of 0.4 ammonium sulfate and 3.75 ml of 1.0 acetate buffer. Each extraction should be centrifuged and separated as before, discarding the ether layer and retaining the aqueous layer.

The aqueous ammonium sulfate extracts are shaken individually with about 38 ml of ether. Allow to stand about 30-45 min, separate and discard the ether layer. Filter the aqueous layer through a medium porosity sintered glass funnel with 0.5 in. of Celite or other filter aid. Approximately 450 ml of filtrate should be recovered and should be clear yellow. Dialyze the filtrate in Visking tubing (0.75 in.) overnight in 750 ml of saturated ammonium sulfate which has been adjusted to pH 4 with glacial acetic acid. The ammonium sulfate solution should be stirred with a magnetic stirrer. Repeat the dialysis with 750 ml of fresh saturated ammonium sulfate.

The phosvitin precipitates as a white gelatinous precipitate and is separated by centrifugation. Disperse the precipitate in 100 ml of 0.5 *M* sodium chloride, dialyze with several (2 to 3) changes of 2 *M* sodium chloride and then in distilled water. Add 5 drops of toluene to the contents of the dialysis sacks as a preservative. The dialysis sacks should not be more than half filled since they will absorb water as the concentration of salt decreases.

Concentrate the contents of the dialysis sacks in an air stream at 4° C then finally by lyophilization. Record the yield.

Product Examination.—Determine the percent of phosphorous (as phosphate) and nitrogen. Hydrolysis followed by Kjeldahl or Nessler's technique can be used for determining the percent nitrogen. Hydrolysis to release the phosphate which can then be determined by a modified Fisk-Subbarow method is generally used for phosphate analysis. The ratio of N/P should be about 2.7-2.8.

QUESTIONS

(1) Why must the phosvitin be hydrolyzed before phosphate analysis?

(2) Suggest an alternative procedure whereby the dialysis procedures might be eliminated.

Appendix A

PREPARATION OF SOLUTIONS

Preparation of *M*/64 Sucrose

Sucrose (mol. wt. 342.30) To prepare *M*/64, dissolve 5.348 g sucrose in 1000 ml distilled water.

Preparation of *N*/10 Sodium hydroxide mol. wt. 40.00)

Solid sodium hydroxide is always contaminated with carbonate. Consequently a saturated solution of sodium hydroxide is frequently employed for the preparation of 0.1 *N* sodium hydroxide solution. Carbonates are not soluble and precipitate out in a saturated solution thus leaving a carbonate-free solution. Furthermore, once the concentration of the saturated solution has been determined, it may be utilized in the preparation of solutions of lesser concentrations. Mix about 110 g sodium hydroxide with 100 ml of distilled water in a 300 ml Erlenmeyer flask (Pyrex). Stopper and allow to stand for a couple of days or until the sodium carbonate precipitates, leaving a clear solution of sodium hydroxide.

A saturated solution of sodium hydroxide has a specific gravity of 1.53 and is about a 50.1% solution. A 100 ml of the saturated solution will contain 1.53 g $\times$ 50.1 or 76.65 g of sodium hydroxide per 100 ml (or .7665 g/ml). A liter of 0.1 *N* sodium hydroxide contains 4.0 g of sodium hydroxide. Since a milliliter of concentrated sodium hydroxide contains 0.7665 g, then $4.0 \div 0.7665 = 5.2$, or the number of milliliters of saturated sodium hydroxide which will contain 4.0 g of sodium hydroxide, and which when diluted to 1 liter will make an approximately 0.1 *N* solution. Measure out 5.2 ml of the saturated sodium hydroxide solution and dilute to 1 liter with carbon dioxide-free distilled water. Standardize by titration against a standard acid or against acid potassium phthalate, a primary standard.

Standardization by Titration Against a Primary Standard (Acid Potassium Phthalate)

Acid potassium phthalate (mol. wt. 204.2) contains one replaceable hydrogen and, as a result, its molecular weight is its equivalent weight. Dry a small amount of pure acid potassium phthalate in a flat type weighing bottle for 2 hr or longer at 110°-120°C and cool in a desiccator. Weigh accurately about 0.5 g sample of the dried primary standard into a 300 ml Erlenmeyer flask. Dissolve in 50 ml of freshly boiled and cooled water. Titrate the acid potassium phthalate solution with the sodium hydroxide solution to be standardized using phenophthalien as an indicator. The first noticeable persistent pink coloration is the endpoint.

Preparation of Some Common Indicator Solutions for Neutralization Reactions

Bromthymol Blue: Dissolve 0.1 gm of the indicator in 100 ml of 50% ethanol solution.

Bromcresol Purple: Grind together in an agate mortar 0.100 g of brocresol purple with 9 ml of 0.02 *N* sodium hydroxide solution. Dilute with water to 200 ml and filter if necessary.

Congo Red: Dissolve 0.5 g of congo red in 100 ml of a 10% ethanol solution.

Methyl orange: Dissolve 0.1 g of methyl orange in 100 ml of hot water. Filter as solution becomes cloudy on standing.

Methyl Red: Dissolve 0.2 g of methyl red in 60 ml of 95% ethanol and dilute to 100 ml with water.

Phenolphthalein: Dissolve 1 g of phenolphthalein in 50 ml of 95% ethanol and add 50 ml of water.

Thymol Blue: Grind in an agate mortar 0.10 g of thymol blue with 21.5 ml of 0.10 *N* sodium hydroxide and dilute to 250 ml with water.

Preparation of 0.1 *N* Hydrochloric Acid

Concentrated hydrochloric acid is about 12 *N* or 38.0% HCl weight in volume. Approximately 0.1 *N* HCl may be prepared by diluting 9 ml of the concentrated acid to 1 liter in a volumetric flask. This must be standardized by titration against a *N*/10 standard base, using methyl red as an indicator or against a primary standard (sodium carbonate or Borax).

Standardization of Acid by Titration Against a Primary Standard (Sodium Carbonate)

Dry a shallow layer of analytical grade sodium carbonate in a weighing bottle for 2 hr at 130°-150°C, then cover and cool in a desiccator. Weigh accurately, by difference, several 0.1-0.2 g samples of the above sodium carbonate and dissolve each in 100 ml of distilled water.

Add 2-3 drops of methyl orange to the sodium carbonate solution, and, with constant stirring, add the acid solution from a burette until the endpoint is reached. Carefully wash down the sides of the beaker with the wash bottle. If the solution changes color, add more acid until the endpoint is obtained.

From the weight of sodium carbonate sample and the milliliters of acid required to neutralize it, calculate the normality factor (NF) of the acid solution.

$$\text{ml} \times \text{NF} \times \text{Meq of } Na_2CO_3\left(\frac{106.0}{2000}\right) = \text{Wt of } Na_2CO_3$$

Replicate normality factors should check within 5 in fourth decimal place to be acceptable.

Standardization of Acid by Titration Against a Primary Standard—Borax ($Na_2 B_4 0_7 .10H_2$)

Dissolve 15 g of analyzed c.p. grade Borax in 50 ml of water with heating. On cooling, the decahydrate crystallizes out below 55°C and is separated from the liquor by filtering through filter paper using a Büchner funnel and suction. Wash the crystals twice with water, followed twice with 5 ml portions of ethyl alcohol, then twice with 5 ml portion of ethyl portion of ethyl ether. Spread the crystals in a thin layer on a watch glass at room temperature until the crystals are dry. The dry crystals may be stored in a tightly closed container for reasonable periods of time.

Accurately weigh samples of approximately 0.5 g of the above prepared sodium tetraborate, dissolve in 25 ml of distilled water, add 5 drops of methyl red indicator and titrate to the endpoint. From the weight of sample and the milliliters of acid required to neutralize it, calculate the normality factor of the acid.

Oxidation-Reduction Reactions

Numerous methods are based upon chemical reactions involving valence changes rather than the interaction of H^+ cations and OH^- anions of neutralization and precipitation methods. It will be recalled that oxidation is a gain in positive charge and reduction is a gain in negative charge. Expressed another way: Oxidation is the result of a loss of one or more electrons by an atom, radical, or ion. Conversely, reduction is a gain of one or more electrons by an atom, radical or ion.

Balancing of Oxidation-Reduction Equations

In writing and balancing oxidation-reduction equations, it is necessary to know the substances involved in the reaction and the products formed. The latter kind of information may be obtained by experiment or from textbooks. The next step involves writing the equation in an unbalanced molecular form. For example, the unbalanced equation for the oxidation of ferrous sulfate by potassium permanganate, in acid solution, would be:

$$FeSO_4 + KMnO_4 + H_2SO_4 \Rightarrow Fe_2(SO_4)_3 + MnSO_4 + K_2SO_4 + H_2O$$

The electron transfer for the ions concerned in the oxidation-reduction would be:

-1e- ($Fe^{+2} \rightarrow Fe^{+3}$)

$$[MnO_4]^{-1} + Fe^{+2} \rightarrow Mn^{+2} + Fe^{+3}$$

+5e- ($Mn^{+7} \rightarrow Mn^{+2}$)

Mn^{+7} Mn^{+2}

The electrons gained by the manganese atom are lost by the ferrous atom in being oxidized to the ferric state. Since five electrons are gained by the manganese atom, five electrons must be lost by the iron; thus each iron atom must be multiplied by five, i.e., the number of electrons lost by the reducing agent must be equivalent to the number of electrons gained by the oxidizing agent.

Rewriting the above equation:

-1e- × 5 ($5Fe^{+2} \rightarrow 5Fe^{+3}$)

$$[MnO_4]^{-1} + 5Fe^{+2} \rightarrow Mn^{+2} + 5Fe^{+3}$$

+5e- × 1 ($Mn^{+7} \rightarrow Mn^{+2}$)

Mn^{+7} Mn^{+2}

On completing the equation with the other atoms involved, but which do not undergo electron transfers, it is seen that at least two molecules of potassium permanganate will be required to produce one molecule of potassium sulfate. This necessitates multiplying all coefficients by two, and when this is done the equation is:

$$2\ KMnO_4 + 10\ FeSO_4 + H_2SO_4 \Rightarrow 5\ Fe_2(SO_4) + 2\ MnSO_4 + K_2SO_4 + H_2O$$

Note, eighteen SO_4 radicals are present on the right-hand side of the equation, and thus eight H_2SO_4 molecules must be required on the left hand side to balance the equation. The balanced equation is written

$$2\ KMnO_4 + 10\ FeSO_4 + 8\ H_2SO_4 \Rightarrow 5\ Fe_2(SO_4)_3 + 2\ MnSO_4 + K_2SO_4 + 8\ H_2O$$

Preparation of an Approximately *N*/10 Potassium Permanganate Solution

It will be recalled that a normal solution contains one gram-equivalent weight of the oxidizing substance per liter (8.00 g oxygen) which corresponds to one unit change in valence. In an acid solution for $KMnO_4$ we have:

$$Mn^{+7} \Rightarrow Mn^{+2}$$

This represents a change in valence of five or a net increase of five positive charges (gain in electrons) per manganese atom. With these facts in mind, a normal solution would contain one-fifth of a mole of potassium permanganate or 31.61 g. A 0.1 *N* solution of $KMnO_4$ would contain 3.161 g per liter of solution.

Dissolve 3.20 g of reagent grade potassium permanganate in a liter of distilled water and boil for 10-15 min. Allow the solution to stand overnight; filter through a filtering crucible, and store in a clean brown glass bottle with a glass stopper. The filtering removes all particles of manganese dioxide which, if allowed to remain in the solution, would catalyze further decomposition of the solution. Similarly, filter paper should not be used in filtering, since particles of the paper suspended in the solution would lead to further reduction. The solution should be kept in the dark except when in use. Subsequent decomposition of the solution can be recognized by the appearance of a coating on the walls of the bottle.

Standardization of Potassium Permanganate by Titration Against a Primary Standard—Sodium Oxalate ($Na_2C_2O_4$)

The reactions involved when sodium oxalate is standardized by titration with potassium permanganate are:

$$5\ Na_2C_2O_4 + 5\ H_2SO_4 \Rightarrow 5\ Na_2SO_4 + 5\ H_2C_2O_4$$

$$5\ H_2C_2O_4 + 3\ H_2SO_4 + 2\ KMnO_4 \Rightarrow K_2SO_4 + 2\ MnSO_4 + 10\ CO_2 + 8\ H_2O$$

Inasmuch as five moles of sodium oxalate react with 10 equivalent weights of potassium permanganate, one equivalent weight of potassium permanganate will react with one-half mole of sodium oxalate. The equivalent weight of sodium oxalate is 134.00/2 or 67.00 g and the milliequivalent weight is 0.0670 g.

Weight accurately, by difference, three 0.25 g samples of dry $Na_2C_2O_4$, and transfer to 400 ml beakers. Prepare approximately 1.5 *N* sulfuric acid solution by pouring 13 ml concentrated H_2SO_4 into 300 ml of water. Dissolve each oxalate sample in 75 ml 1.5 *N* sulfuric acid solution.

Heat the solution to 80-90°C, remove from the flame, and titrate this solution at once with the permanganate solution to be standardized, stirring vigorously throughout the titration using a thermometer for a stirring rod. The endpoint is a permanent pink color. The temperature should not fall below 60° during the titration. The reaction is catalyzed by manganous ions; the first portions of permanganate added react slowly, but, after some manganous ion is formed, succeeding portions of permanganate are decolorized almost instantly. Obtain a blank by heating 75 ml 1.5 *N* sulfuric acid to 80° and titrating with permanganate.

Calculate the normal factor of potassium permanganate from the corrected volume of permanganate and the weight of the primary standard. Replicate normal factors should not exceed 4-5 parts per 1000.

$$\text{ml} \times \text{Normal factor} \times \frac{Na_2C_2O_4}{2000} = \text{g } Na_2C_2O_4$$

Biuret Reagent (Quantitative)

Dissolve 0.75 g of $CuSO_4 \cdot 5H_2O$ and 3.0 g of $NaKC_4H_4O_6 \cdot 4H_2O$ (sodium potassium tartrate) in approximately 250 ml of water in a 500 ml volumetric flask. Add with stirring 150 ml of freshly prepared carbonate-free 10% NaOH. Make up to 500 ml with water and store in a plastic bottle. (A red or black precipitate indicates that the solution is no longer useable.)

Ninhydrin Reagent (Quantitative)

Dissolve 200 mg of $SnCl_2 \cdot 2H_2O$ in 125 ml of 0.2M citrate buffer, pH 5. Add this solution to 125 ml of 1,4-dioxane containing 5 g of dissolved ninhydrin. Store in refrigerator.

TABLE A.1
TABLE FOR TEMPERATURE CORRECTION OF READING OF THE REFRACTOMETER

Temperature °C		Reading														
		0	5	10	15	20	25	30	35	40	45	50	55	60	65	70
10	Deduct from reading	0.50	0.54	0.58	0.61	0.64	0.66	0.68	0.70	0.72	0.73	0.74	0.75	0.76	0.78	0.79
11		0.46	0.46	0.53	0.55	0.58	0.60	0.62	0.64	0.65	0.66	0.67	0.68	0.64	0.70	0.71
12		0.42	0.45	0.48	0.50	0.52	0.54	0.56	0.57	0.58	0.59	0.60	0.61	0.61	0.63	0.63
13		0.37	0.40	0.42	0.44	0.46	0.48	0.49	0.50	0.51	0.52	0.53	0.54	0.54	0.55	0.55
14		0.33	0.35	0.37	0.39	0.40	0.41	0.42	0.43	0.44	0.45	0.45	0.46	0.46	0.47	0.48
15		0.27	0.29	0.31	0.33	0.34	0.34	0.35	0.36	0.37	0.37	0.38	0.39	0.39	0.40	0.40
16		0.22	0.24	0.25	0.26	0.27	0.28	0.28	0.29	0.30	0.30	0.30	0.31	0.31	0.32	0.32
17		0.17	0.18	0.19	0.20	0.21	0.22	0.21	0.22	0.22	0.23	0.23	0.23	0.23	0.24	0.24
18		0.12	0.13	0.13	0.14	0.14	0.14	0.11	0.15	0.15	0.15	0.15	0.16	0.16	0.16	0.16
19		0.06	0.06	0.06	0.07	0.07	0.07	0.07	0.08	0.08	0.08	0.08	0.08	0.08	0.08	0.08
20		0	0	0	0	0	0	0	0	0	0	0	0	0	0	0
21	Add to reading	0.06	0.07	0.07	0.07	0.07	0.08	0.08	0.08	0.08	0.08	0.08	0.08	0.08	0.08	0.08
22		0.13	0.13	0.14	0.14	0.15	0.15	0.15	0.15	0.15	0.16	0.16	0.16	0.16	0.16	0.16
23		0.19	0.20	0.21	0.22	0.22	0.23	0.23	0.23	0.23	0.24	0.24	0.24	0.24	0.24	0.24
24		0.26	0.27	0.28	0.29	0.30	0.30	0.31	0.31	0.31	0.31	0.31	0.32	0.32	0.32	0.32
25		0.33	0.35	0.36	0.37	0.38	0.38	0.39	0.39	0.40	0.40	0.40	0.40	0.40	0.40	0.40
26		0.40	0.42	0.43	0.44	0.45	0.46	0.47	0.47	0.48	0.48	0.48	0.48	0.48	0.48	0.48
27		0.48	0.50	0.52	0.53	0.54	0.55	0.55	0.55	0.56	0.56	0.56	0.56	0.56	0.56	0.56
28		0.56	0.57	0.60	0.61	0.62	0.63	0.63	0.63	0.64	0.64	0.64	0.64	0.64	0.64	0.64
29		0.64	0.66	0.68	0.69	0.71	0.72	0.72	0.72	0.73	0.73	0.73	0.73	0.73	0.73	0.73
30		0.72	0.74	0.77	0.78	0.79	0.80	0.80	0.81	0.81	0.81	0.81	0.81	0.81	0.81	0.81

This table was prepared under the authorization of International Sugar Analysis Control Committee (1949).

TABLE A.2
RELATION BETWEEN READING OF THE REFRACTOMETER AND REFRACTIVE INDEX (^nD) (20°C)

Reading (%)	(^nD)	Reading (%)	(^nD)	Reading (%)	(^nD)	Reading (%)	(^nD)
0	1.33299	19	1.36218	38	1.3958	57	1.4351
1	1.33423	20	1.36384	39	1.3987	58	1.4373
2	1.33588	21	1.36551	40	1.3997	59	1.4396
3	1.33733	22	1.36719	41	1.4016	60	1.4418
4	1.33880	23	1.36888	42	1.4036	61	1.4441
5	1.34027	24	1.37059	43	1.4056	62	1.4464
6	1.34176	25	1.3723	44	1.4076	63	1.4486
7	1.34326	26	1.3740	45	1.4096	64	1.4509
8	1.34477	27	1.3758	46	1.4117	65	1.4532
9	1.34629	28	1.3775	47	1.4137	66	1.4555
10	1.34783	29	1.3793	48	1.4158	67	1.4579
11	1.34937	30	1.3811	49	1.4179	68	1.4603
12	1.35093	31	1.3829	50	1.4200	69	1.4627
13	1.35250	32	1.3847	51	1.4221	70	1.4651
14	1.35408	33	1.3865	52	1.4242	71	1.4676
15	1.35567	34	1.3883	53	1.4264	72	1.4700
16	1.35728	35	1.3902	54	1.4285	73	1.4725
17	1.35890	36	1.3920	55	1.4307	74	1.4749
18	1.36053	37	1.3939	56	1.4329	75	1.4774

TABLE A.3
COMPOSITION OF SOME INORGANIC ACIDS AND BASES

	Molecular Weight	Normality	*M* (Molarity)	Specific Gravity	Concentration (%)
Acetic acid	60.05	17.5	17.5	1.05	99–100
Ammonium hydroxide	35.05	7.4	7.4	0.90	28–30
Hydrochloric acid	36.46	12.0	12.0	1.18	36.5–38.0
Phosphoric acid	98.00	44.1	14.7	1.7	85
Sulfuric acid	98.08	36.0	18.0	1.84	95–98
Nitric acid	63.0	16.0	16.0	1.42	70

TABLE A.4
LOGARITHMS

Natural Numbers	0	1	2	3	4	5	6	7	8	9	PP 1	PP 2	PP 3	PP 4	PP 5	PP 6	PP 7	PP 8	PP 9
10	0000	0043	0086	0128	0170	0212	0253	0294	0334	0374	4	8	12	17	21	25	29	33	37
11	0414	0453	0492	0531	0569	0607	0645	0682	0719	0755	4	8	11	15	19	23	26	30	34
12	0792	0828	0864	0899	0934	0969	1004	1038	1072	1106	3	7	10	14	17	21	24	28	31
13	1139	1173	1206	1239	1271	1303	1335	1367	1399	1430	3	6	10	13	16	19	23	26	29
14	1461	1492	1523	1553	1584	1614	1644	1673	1703	1732	3	6	9	12	15	18	21	24	27
15	1761	1790	1818	1847	1875	1903	1931	1959	1987	2014	3	6	8	11	14	17	20	22	25
16	2041	2068	2095	2122	2148	2175	2201	2227	2253	2279	3	5	8	11	13	16	18	21	24
17	2304	2330	2355	2380	2405	2430	2455	2480	2504	2529	2	5	7	10	12	15	17	20	22
18	2553	2577	2601	2625	2648	2672	2695	2718	2742	2765	2	5	7	9	12	14	16	19	21
19	2788	2810	2833	2856	2878	2900	2923	2945	2967	2989	2	4	7	9	11	13	16	18	20
20	3010	3032	3054	3075	3096	3118	3139	3160	3181	3201	2	4	6	8	11	13	15	17	19
21	3222	3243	3263	3284	3304	3324	3345	3365	3385	3404	2	4	6	8	10	12	14	16	18
22	3424	3444	3464	3483	3502	3522	3541	3560	3579	3598	2	4	6	8	10	12	14	15	17
23	3617	3636	3655	3674	3692	3711	3729	3747	3766	3784	2	4	6	7	9	11	13	15	17
24	3802	3820	3838	3856	3874	3892	3909	3927	3945	3962	2	4	5	7	9	11	12	14	16
25	3979	3997	4014	4031	4048	4065	4082	4099	4116	4133	2	3	5	7	9	10	12	14	15
26	4150	4166	4183	4200	4216	4232	4249	4265	4281	4298	2	3	5	7	8	10	11	13	15
27	4314	4330	4346	4362	4378	4393	4409	4425	4440	4456	2	3	5	6	8	9	11	13	14
28	4472	4487	4502	4518	4533	4548	4564	4579	4594	4609	2	3	5	6	8	9	11	12	14
29	4624	4639	4654	4669	4683	4896	4713	4728	4742	4757	1	3	4	6	7	9	10	12	13
30	4771	4786	4800	4814	4829	4843	4857	4871	4886	4900	1	3	4	6	7	9	10	11	13
31	4914	4928	4942	4955	4969	4983	4997	5011	5024	5038	1	3	4	6	7	8	10	11	12
32	5051	5065	5079	5092	5105	5119	5132	5145	5159	5172	1	3	4	5	7	8	9	11	12
33	5185	5198	5211	5224	5237	5250	5263	5276	5289	5302	1	3	4	5	6	8	9	10	12
34	5315	5328	5340	5353	5366	5378	5391	5403	5416	5428	1	3	4	5	6	8	9	10	11
35	5441	5453	5465	5478	5490	5502	5514	5527	5539	5551	1	2	4	5	6	7	9	10	11
36	5563	5575	5587	5599	5611	5623	5635	5647	5658	5670	1	2	4	5	6	7	8	10	11
37	5682	5694	5705	5717	5729	5740	5752	5763	5775	5786	1	2	3	5	6	7	8	9	10
38	5798	5809	5821	5832	5843	5855	5866	5877	5888	5899	1	2	3	5	6	7	8	9	10
39	5911	5922	5933	5944	5955	5966	5977	5988	5999	6010	1	2	3	4	5	7	8	9	10
40	6021	6031	6042	6053	6064	6075	6085	6096	6107	6117	1	2	3	4	5	6	8	9	10
41	6128	6138	6149	6160	6170	6180	6191	6201	6212	6222	1	2	3	4	5	6	7	8	9
42	6232	6243	6253	6263	6274	6284	6294	6304	6314	6325	1	2	3	4	5	6	7	8	9
43	6335	6345	6355	6365	6375	6385	6395	6405	6415	6425	1	2	3	4	5	6	7	8	9
44	6435	6444	6454	6464	6474	6484	6493	6503	6513	6522	1	2	3	4	5	6	7	8	9
45	6532	6542	6551	6561	6571	6580	6590	6599	6609	6618	1	2	3	4	5	6	7	8	9
46	6628	6637	6646	6655	6665	6675	6684	6693	6702	6712	1	2	3	4	5	6	7	7	8
47	6721	6730	6739	6749	6758	6767	6776	6785	6794	6803	1	2	3	4	5	5	6	7	8
48	6812	6821	6830	6839	6848	6857	6866	6875	6884	6893	1	2	3	4	4	5	6	7	8
49	6902	6911	6920	6928	6937	6946	6955	6964	6972	6981	1	2	3	4	4	5	6	7	8
50	6990	6998	7007	7016	6024	7033	7042	7050	7059	7067	1	2	3	3	4	5	6	7	8
51	7076	7084	7093	7101	7110	7118	7126	7135	7143	7152	1	2	3	3	4	5	6	7	8
52	7160	7168	7177	7185	7193	7202	7210	7218	7226	7235	1	2	2	3	4	5	6	7	7
53	7243	7251	7259	7267	7275	7284	7292	7300	7308	7316	1	2	2	3	4	5	6	6	7
54	7324	7332	7340	7348	7356	7364	7372	7380	7388	7396	1	2	2	3	4	5	6	6	7

Natural Numbers	0	1	2	3	4	5	6	7	8	9	PP 1	PP 2	PP 3	PP 4	PP 5	PP 6	PP 7	PP 8	PP 9
55	7404	7412	7419	7427	7435	7443	7451	7459	7466	7474	1	2	2	3	4	5	5	6	7
56	7482	7490	7497	7505	7513	7520	7528	7536	7543	7551	1	2	2	3	4	5	5	6	7
57	7559	7566	7574	7582	7589	7597	7604	7612	7619	7627	1	2	2	3	4	5	5	6	7
58	7634	7642	7649	7657	7664	7672	7679	7680	7694	7701	1	1	2	3	4	4	5	6	7
59	7709	7716	7723	7731	7738	7745	7752	7760	7767	7774	1	1	2	3	4	4	5	6	7
60	7782	7789	7796	7803	7810	7818	7825	7832	7839	7846	1	1	2	3	4	4	5	6	6
61	7853	7860	7868	7875	7882	7889	7896	7903	7910	7917	1	1	2	3	4	4	5	6	6
62	7924	7931	7938	7945	7952	7959	7966	7973	7980	7987	1	1	2	3	3	4	5	6	6
63	7993	8000	8007	8014	8021	8028	8035	8041	8048	8055	1	1	2	3	3	4	5	5	6
64	8062	8069	8075	8082	8089	8096	8102	8109	8116	8122	1	1	2	3	3	4	5	5	6
65	8129	8136	8142	8149	8156	8162	8169	8176	8182	8189	1	1	2	3	3	4	5	5	6
66	8195	8202	8209	8215	8222	8228	8235	8241	8248	8254	1	1	2	3	3	4	5	5	6
67	8261	8267	8274	8280	8287	8293	8299	8306	8312	8319	1	1	2	3	3	4	5	5	6
68	8325	8331	8338	8344	8351	8357	8363	8370	8376	8382	1	1	2	3	3	4	4	5	6
69	8388	8395	8401	8407	8414	8420	8426	8432	8439	8445	1	1	2	2	3	4	4	5	6
70	8451	8457	8463	8470	8476	8482	8488	8494	8500	8506	1	1	2	2	3	4	4	5	6
71	8513	8519	8525	8531	8537	8543	8549	8555	8561	8567	1	1	2	2	3	4	4	5	5
72	8573	8579	8585	8591	8597	8603	8609	8615	8621	8627	1	1	2	2	3	4	4	5	5
73	8633	8639	8645	8651	8657	8663	8669	8675	8681	8686	1	1	2	2	3	4	4	5	5
74	8692	8698	8704	8710	8716	8722	8727	8733	8739	8745	1	1	2	2	3	4	4	5	5
75	8751	8756	8762	8768	8774	8779	8785	8791	8798	8802	1	1	2	2	3	3	4	5	5
76	8908	8814	8820	8825	8831	8837	8842	8848	8854	8859	1	1	2	2	3	3	4	5	5
77	8865	8871	8876	8882	8887	8893	8899	8904	8910	8915	1	1	2	2	3	3	4	4	5
78	8921	8927	8932	8938	8943	8949	8954	8960	8965	8971	1	1	2	2	3	3	4	4	5
79	8976	8982	8987	8993	8998	9004	9009	9015	9020	9026	1	1	2	2	3	3	4	4	5
80	9031	9036	9042	9047	9053	9058	9063	9069	9074	9079	1	1	2	2	3	3	4	4	5
81	9085	9090	9096	9101	9106	9112	9117	9122	9128	9133	1	1	2	2	3	3	4	4	5
82	9138	9143	9149	9154	9159	9165	9170	9175	9180	9186	1	1	2	2	3	3	4	4	5
83	9191	9196	9201	9206	9212	9217	9222	9227	9232	9238	1	1	2	2	3	3	4	4	5
84	9243	9248	9253	9258	9263	9269	9274	9279	9284	9289	1	1	2	2	3	3	4	4	5
85	9294	9299	9304	9309	9315	9320	9325	9330	9335	9340	1	1	2	2	3	3	4	4	5
86	9345	9350	9355	9360	9365	9370	9375	9380	9385	8390	1	1	2	2	3	3	4	4	5
87	9395	9400	9405	9410	9415	9420	9425	9430	9435	9440	0	1	1	2	2	3	3	4	4
88	9445	9450	9455	9460	9465	9469	9474	9479	9484	9489	0	1	1	2	2	3	3	4	4
89	9494	9499	9504	9509	9513	9518	9523	9528	9533	9538	0	1	1	2	2	3	3	4	4
90	9542	9547	9552	9557	9562	9566	9571	9576	9581	9586	0	1	1	2	2	3	3	4	4
91	9590	9595	9600	9605	9609	9614	9619	9624	9628	9633	0	1	1	2	2	3	3	4	4
92	9638	9643	9647	9652	9657	9661	9666	9671	9675	9680	0	1	1	2	2	3	3	4	4
93	9685	9689	9694	9699	9703	9708	9713	9717	9722	9727	0	1	1	2	2	3	3	4	4
94	9731	9736	9741	9745	9750	9754	9759	9763	9768	9773	0	1	1	2	2	3	3	4	4
95	9777	9782	9786	9791	9795	9800	9805	9809	9814	9818	0	1	1	2	2	3	3	4	4
96	9823	9827	9832	9836	9841	9845	9850	9854	9859	9863	0	1	1	2	2	3	3	4	4
97	9868	9872	9877	9881	9886	9890	9894	9899	9903	9908	0	1	1	2	2	3	3	4	4
98	9912	9917	9921	9926	9930	9934	9939	9943	9948	9952	0	1	1	2	2	3	3	4	4
99	9956	9961	9965	9969	9974	9978	9983	9987	9991	9996	0	1	1	2	2	3	3	3	4

TABLE A.5
TABLE OF ATOMIC WEIGHTS

Based on the Atomic Mass of $^{12}C=12$

Name	Symbol	Atomic weight	Name	Symbol	Atomic weight
Actinium	Ac	...	Mercury	Hg	200.59
Aluminum	Al	26.9815	Molybdenum	Mo	95.94
Americium	Am	...	Neodymium	Nd	144.24
Antimony	Sb	121.75	Neon	Ne	20.183
Argon	Ar	39.948	Neptunium	Np	...
Arsenic	As	74.9216	Nickel	Ni	58.71
Astatine	At	...	Niobium	Nb	92.906
Barium	Ba	137.34	Nitrogen	N	14.0067
Berkelium	Bk	...	Nobelium	No	...
Beryllium	Be	9.0122	Osmium	Os	190.2
Bismuth	Bi	208.980	Oxygen	O	15.9994
Boron	B	10.811	Palladium	Pd	106.4
Bromine	Br	79.909	Phosphorus	P	30.9738
Cadmium	Cd	112.40	Platinum	Pt	195.09
Calcium	Ca	40.08	Plutonium	Pu	...
Californium	Cf	...	Polonium	Po	...
Carbon	C	12.01115	Potassium	K	39.102
Cerium	Ce	140.12	Praseodymium	Pr	140.907
Cesium	Cs	132.905	Promethium	Pm	...
Chlorine	Cl	35.453	Protactinium	Pa	...
Chromium	Cr	51.996	Radium	Ra	...
Cobalt	Co	58.9332	Radon	Rn	...
Copper	Cu	63.54	Rhenium	Re	186.2
Curium	Cm	...	Rhodium	Rh	102.905
Dysprosium	Dy	162.50	Rubidium	Rb	85.47
Einsteinium	Es	...	Ruthenium	Ru	101.07
Erbium	Er	167.26	Samarium	Sm	150.35
Europium	Eu	151.96	Scandium	Sc	44.956
Fermium	Fm	...	Selenium	Se	78.96
Fluorine	F	18.9984	Silicon	Si	28.086
Francium	Fr	...	Silver	Ag	107.870
Gadolinium	Gd	157.25	Sodium	Na	22.9898
Gallium	Ga	69.72	Strontium	Sr	87.62
Germanium	Ge	72.59	Sulfur	S	32.064
Gold	Au	196.967	Tantalum	Ta	180.948
Hafnium	Hf	178.49	Technetium	Tc	...
Helium	He	4.0026	Tellurium	Te	127.60
Holmium	Ho	164.930	Terbium	Tb	158.924
Hydrogen	H	1.00797	Thallium	Tl	204.37
Indium	In	114.82	Thorium	Th	232.038
Iodine	I	126.9044	Thulium	Tm	168.934
Iridium	Ir	192.2	Tin	Sn	118.69
Iron	Fe	55.847	Titanium	Ti	47.90
Krypton	Kr	83.80	Tungsten	W	183.85
Lanthanum	La	138.91	Uranium	U	238.03
Lead	Pb	207.19	Vanadium	V	50.942
Lithium	Li	6.939	Xenon	Xe	131.30
Lutetium	Lu	174.97	Ytterbium	Yb	173.04
Magnesium	Mg	24.312	Yttrium	Y	88.905
Manganese	Mn	54.9380	Zinc	Zn	65.37
Mendelevium	Md	...	Zirconium	Zr	91.22

Appendix B

CALCULATIONS USED IN FOOD ANALYSIS[1]

Abstract

The more common calculations used in the analysis of food in one government laboratory are given with examples. Equations are set up so that data can be fed into them. The examples serve as checks for calculations and whether meaningful values have been obtained. Factors and significant figures extend the usefulness of the calculations.

Introduction

One of the problems in orienting new chemists in a food testing laboratory is that of acquainting them with the calculations commonly used. If they are provided with equations into which they can feed data, and examples making use of actual test data, almost immediate use can be made of their work. For they not only are given a valuable working tool, but they also gain a much more rapid and clearer insight into methods of analysis, what they are to expect of the methods, and what is expected of them as analysts.

The calculations shown here are some of the more common ones used with the greatest frequency in this laboratory. For the most part, they are based on methods of analysis found in the three major methods references (1, 2, 3) which have particular application in the United States of America Government's analysis of foods and food products. The equations are set up for the express purpose of feeding data into them with the expectation of performing calculations correctly.

The specific examples are the result of actual analyses. The values shown as answers are in line with current government specification requirements. They represent desirable levels in good quality foods, and may be used as indications of the reasonableness of analytical data. The number of significant figures shown in an answer is indicative of the precision expected of the answer for the particular analysis. Likewise, the number of significant figures shown for sample weights, volumes, and normalities is indicative of the number of significant figures that should be achieved in practice.

The usefulness of the calculations is extended by the use of factors. For example, the calculation for "PROTEIN, PERCENT," gives factors for converting percent nitrogen into percent protein in a wide variety of foods. The factors were obtained from the references.

References

1. CEREAL LABORATORY METHODS, American Association of Cereal Chemists, Inc., 1955 University Avenue, St. Paul, Minnesota 55104.

2. OFFICIAL AND TENTATIVE METHODS OF THE AMERICAN OIL CHEMISTS' SOCIETY, American Oil Chemists' Society, 35 East Wacker Drive, Chicago, Illinois 60601.

3. OFFICIAL METHODS OF ANALYSIS OF THE ASSOCIATION OF OFFICIAL AGRICULTURAL CHEMISTS, Association of Official Analytical Chemists, P.O. Box 540, Benjamin Franklin Station, Washington, D.C. 20044.

ACIDITY, PERCENT:

$$\frac{(\text{ml base})\ (N\ \text{base})\ (\text{meq* wt acid})\ (100)}{(\text{g sample})} = \text{percent acid}$$

*NOTE: meq = milliequivalent.
Meq wt acetic acid = 0.06005
Meq wt citric acid = 0.06404

[1] By Charles H. Coleman, Defense Subsistence Testing Laboratory, 1819 West Pershing Road, Chicago, Illinois.

EXAMPLE:

$$\frac{(25.20 \text{ ml NaOH})\ (0.1000\ N)(0.06005 \text{ meq wt acetic acid})\ (100)}{(3.000 \text{ g vinegar})} = 5.0\% \text{ acetic acid}$$

ACIDS, FREE FATTY, PERCENT:

$$\frac{(\text{ml base})\ (N \text{ base})(\text{meq wt free fatty acid*})}{\text{g sample}} = \text{percent free fatty acids}$$

*NOTE: meq wt oleic acid = 0.282, lauric acid = 0.200, palmitic acid = 0.256.

EXAMPLE:

$$\frac{(10.00 \text{ ml NaOH})\ (0.0100\ N)\ (0.282 \text{ meq wt oleic acid})\ (100)}{(28.2 \text{ g sample})} = 0.10\% \text{ oleic acid (free fatty acid)}$$

ADDED MOISTURE IN SAUSAGE:

$$\frac{(W) - (4P)}{[1 - (0.01W)] + (0.04P)} = \text{percent added moisture}$$

W = percent moisture, P = percent protein = (6.25) (%N). Correct if necessary for protein in added material.

EXAMPLE:

$$\frac{(60\% \text{ moisture}) - [(4)\ (14\% \text{ protein})]}{1 - [(0.01)\ (60\% \text{ moisture})] + [(0.04)\ (14\% \text{ protein})]} = 4.2\% \text{ added moisture}$$

ALIQUOTS, USING IN CALCULATIONS:
See "DILUTIONS, USING IN CALCULATIONS"

ASH, PERCENT:

$$\frac{(\text{grams ash})\ (100)}{(\text{grams sample})} = \text{percent ash}$$

EXAMPLE:

$$\frac{(0.0700 \text{ g ash})\ (100)}{(5.000 \text{ g pepper})} = 1.4\% \text{ ash}$$

CAFFEINE, BAILEY-ANDREW METHOD:

$$\frac{(\text{ml acid})\ (N \text{ acid})\ (0.014 \text{ meq wt N})\ (3.464^*)\ (\text{dilution})\ (100)}{(\text{g sample})\ (\text{aliquot})} = \text{percent anhydrous caffeine}$$

*NOTE: 3.464 converts nitrogen to caffeine.

EXAMPLE:

$$\frac{(15.26 \text{ ml})\ (0.1000\ N \text{ HCl})\ (0.014)\ (3.464)\ (500 \text{ ml})\ (100)}{(5.000 \text{ g instant coffee})\ (200 \text{ ml})} = 3.7\% \text{ anhydrous caffeine}$$

CALORIES:

Calories = [(9) (g fat) + (4) (g protein) + (4) (g carbohydrate)]

EXAMPLE: Using a sample of 100 g of enriched white bread made with 5-6% nonfat dry milk, and analyzing 35.0% moisture, 3.8% fat, 9.0% protein, 2.0% ash, and 50.2% carbohydrate by difference:

Calories = [(9) (3.8) + (4) (9.0) + (4)(50.2)] = 271 calories

NOTE: The factors of 9,4,4 are general; more exact factors are available in *COMPOSITION OF FOODS*, Agriculture Handbook No. 8, Agricultural Research Service, United States Department of Agriculture, Washington, D.C. 20402, Revised December 1963, page 160, TABLE 6. This reference gives a value of 275 calories for the food used in the above calculation.

CARBOHYDRATE BY DIFFERENCE:

% Carbohydrate = (100%) - [(% moisture) + (% fat) + (% protein) + (% ash)]

EXAMPLE:

% Carbohydrate, enriched white bread (5-6% nonfat dry milk)
= (100%) - [(35.0% moisture) + (3.8% fat) + (9.0% protein) + (2.0% ash)] = 50.2%

COLORIMETRY:

$$\frac{(C_s)\,(OD_x)}{(OD_s)} = C_x$$

C = concentration; *OD* = Optical Density = log 100 - log percent transmission; *s* = standard; *x* = unknown.

CAUTION: OD_s should be adjusted to approximately equal OD_x; otherwise close adherence to Beer's law is necessary.

EXAMPLE:

C_s = 0.94 g ethyl vanillin/100 ml
T_s = 37.5% = transmission of standard. (Log 100 - log 37.5 = 0.426 = OD_s)
T_x = 36.0% = transmission of sample. (Log 100 - log 36.0 = 0.444 = OD_x)

$$\frac{(0.94\text{ g/100 ml})\,(0.444)}{(0.426)} = 0.98\text{ g ethyl vanillin/100 ml sample}$$

DEXTROSE EQUIVALENT:

$$\frac{(\text{g reducing sugar*})\,(100)}{(\text{g total solids})} = \text{DE} = \text{Dextrose Equivalent}$$

*NOTE: Reducing sugar is calculated as dextrose.

EXAMPLE:

$$\frac{(0.4000\text{ g reducing sugar})\,(100)}{(1.000\text{ g total solids})} = 40 = \text{DE}$$

DILUTIONS, USING IN CALCULATIONS:

$$\frac{(\text{g found*})\ (100^{**})\ (\text{volume})}{(\text{g sample})\ (\text{aliquot})} = \text{percent*}$$

*Of characteristic being determined.
**Factor for percent; factor for parts per million = 1×10^6.
NOTE: Volumes and aliquots may be continued indefinitely when used as factors.
EXAMPLE: If 10 ml aliquot of 250 ml original volume is made up to 200 ml, 5 ml aliquot of the 200 used in the determination, the calculation would be:

$$\frac{(0.001\ \text{g found*})\ (100)\ (250\ \text{ml})\ (200\ \text{ml})}{(10\ \text{g sample})\ \ (10\ \text{ml})\ \ (5\ \text{ml})} = 10\%^*$$

DRY BASIS (db):

$$\frac{(\text{percent found*})\ (100)}{(100\% - \text{percent moisture})} = \text{percent* (db)}$$

*Ash, protein, etc.

EXAMPLE:

$$\frac{(0.0294\ \text{g ash})\ (100)\ (100)}{(5.000\ \text{g flour})\ (100\% - 13.5\%\ \text{moisture})} = 0.68\%\ \text{ash (db)}$$

EGG IN DRESSING:

Percent yolk (on a 43% solids basis) = (94.26 P) - (2.192 N)
P = percent of P_2O_5 (*See* "PHOSPHORUS PENTOXIDE" for calculation of P_2O_5); N = percent total nitrogen

EXAMPLE:

$$[(94.26)\ (0.0562\%\ P_2O_5)] - [(2.192)\ (0.137\%\ \text{nitrogen})] = 5.0\%\ \text{egg yolk}$$

EGG, WHOLE, IN ALIMENTARY PASTES:

$$75.5\ \{[(\%\ P_2O_5)\ (1.1)] - (0.055)\} = \text{percent whole egg}$$

See "PHOSPHORUS PENTOXIDE" for calculation of P_2O_5 in unknown. The factor 1.1 makes up for manufacturing loss. The factor 0.055 subtracts lipoid P_2O_5 due to flour.

EXAMPLE:

$$75.5\ \{[(0.171\%\ P_2O_5)\ (1.1)] - (0.055)\} = 10.0\%\ \text{whole egg in noodles}$$

EGG YOLK IN ALIMENTARY PASTES:

$$58\ \{[(\%\ P_2O_5)\ (1.1)] - (0.055)\} = \text{percent egg yolk}$$

See "PHOSPHORUS PENTOXIDE" for calculation of P_2O_5 in unknown. The factor 1.1 makes up for manufacturing loss. The factor 0.055 subtracts lipoid P_2O_5 due to flour.

EXAMPLE:

$$58 \left\{[(0.136\%\ P_2O_5)(1.1)] - (0.055)\right\} = 5.5\% \text{ egg yolk in noodles}$$

FAT OR OIL, PERCENT:

$$\frac{(\text{g found})(100)}{(\text{g sample})} = \text{percent}$$

EXAMPLE:

$$\frac{(0.8000 \text{ g fat})(100)}{(4.000 \text{ g ground beef})} = 20.0\% \text{ fat}$$

FIBER, CRUDE, PERCENT:

$$\frac{(\text{Loss in weight*})(100)}{(\text{g sample})} = \text{percent crude fiber}$$

*Loss in grams due to ashing undigested organic matter.

EXAMPLE:

$$\frac{(0.2500 \text{ g loss})(100)}{(2.500 \text{ g nutmeg})} = 10.0\%$$

IODINE NUMBER:

ml 0.1 $N\ Na_2S_2O_3$ used by blank

- ml 0.1 $N\ Na_2S_2O_3$ used by sample

$$\frac{(\text{ml*})(N\ Na_2S_2O_3)(0.12691 \text{ meq wt iodine})(100)}{(\text{g oil})} = \text{iodine number}$$

*ml $Na_2S_2O_3$ equivalent to I_2 added by sample

EXAMPLE:

$$\frac{(20.49 \text{ ml } Na_2S_2O_3)(0.1000\ N)(0.12691)(100)}{(0.2000 \text{ g paprika oil})} = 130 \text{ iodine number}$$

LANE-EYNON SUGAR METHOD:

$$\frac{(\text{Factor*})(\text{dilution})}{(\text{g sample})(\text{titer})(10^{**})} = \%\ \text{sugar}$$

*From table, Official Methods of Analysis of the AOAC
**Percent multiplier/mg to g factor = 100/1000 = 1/10

EXAMPLE: A 1.5 g sample of syrup is made up to 200 ml and gives a titer of 26.8 ml using 25 ml Lane-Eynon reagent (Soxhlet soln) for a factor of 120.6 for dextrose. The percent dextrose would be calculated as follows:

$$\frac{(120.6)(200 \text{ ml})}{(1.5000 \text{ g syrup})(26.8 \text{ ml})(10)} = 60.0\% \text{ dextrose}$$

MEAT CONTENT:

$$\frac{\%\ \text{meat nitrogen}^* \times 100}{\text{Factor}^{**}} = \%\ \text{lean meat}$$

*NOTE: Nitrogen from added cereals, milk, etc., must be subtracted from total nitrogen to obtain meat nitrogen.

**Factor for various meats = % lean meat/% protein:
4.0 for choice beef = 60/15
4.7 for choice lamb = 77/16.5
4.6 for medium fat pork = 47/10.2
4.4 for medium fat veal = 84/19.1

EXAMPLE: Using sample of carcass beef analyzing 60% lean, 40% fat, 49.4% moisture, 14.9% protein = [(2.384% N) (6.25)]

$$\frac{\%\ \text{lean meat} = 2.384\%\ \text{N} \times 100}{4.0} = 60\%$$

MOISTURE, PERCENT BY DRYING:

$$\frac{(\text{g lost}^*)\ (100)}{(\text{g sample})} = \text{percent moisture}$$

*Empirical method calculates all volatile matter lost during drying as moisture.

EXAMPLE:

$$\frac{(0.2700\ \text{g lost})\ (100)}{(2.0000\ \text{g flour})} = 13.5\%\ \text{moisture}$$

MOISTURE, PERCENT BY TOLUENE DISTILLATION:

$$\frac{(\text{ml water found})\ (\text{sp gr})\ (100)}{(\text{g sample})} = \text{percent moisture}$$

EXAMPLE:

$$\frac{(1.00\ \text{ml water found})\ (1.000\ \text{g/ml})\ (100)}{(50.00\ \text{g dehydrated fish})} = 2.0\%\ \text{moisture}$$

NOTE: The Official Analytical Methods of the American Spice Trade Association do not use a factor for specific gravity; hence, report percent as vol/wt.

MOISTURE, ADDED, IN SAUSAGE:

See "ADDED MOISTURE IN SAUSAGE"

MOISTURE BASIS (mb) 14%:

$$\frac{(\text{percent found}^*)\ (86)}{(100\% - \text{percent moisture})} = \text{percent}^*\ \text{on a 14\% moisture basis}$$

*Ash, protein, etc.

EXAMPLE:

$$\frac{(0.0294 \text{ g ash})\ (100)\ (86)}{(5.000 \text{ g flour})\ (100\% - 13.5\%)} = 0.58\% \text{ ash (14\% moisture basis)}$$

MOISTURE-FREE BASIS (mfb):

See "DRY BASIS (db)"

MOISTURE IN BREAD:

(100% - percent solids*) = % moisture

**See*, "SOLIDS IN BREAD, PERCENT"

EXAMPLE:

(100% - 65% solids) = 35% moisture

OIL OR FAT PERCENT:

See "FAT OR OIL, PERCENT"

OIL, VEGETABLE IN SALAD DRESSING:

(Percent total oil) - [(percent egg yolk) (0.256)] = percent vegetable oil

EXAMPLE:

(41.3% total oil) - [(5.0% egg yolk) (0.256)] = 40.0% vegetable oil

PARTS PER MILLION (ppm):

$$\frac{(\text{g found*})\ (10^6)}{(\text{g sample})} = \text{parts per million (ppm)*}$$

*Of characteristic being determined.
NOTE: Parts per million also = mg/kg = g/1000 kg

EXAMPLE:

$$\frac{(25.00 \text{ ml NaOH})\ (0.0100\ N)\ (0.032 \text{ meq wt } SO_2)\ (10^6)}{(32.00 \text{ g dehydrated potatoes})} = 250 \text{ ppm } SO_2$$

PERCENTAGE OF WEIGHT:

$$\frac{(\text{g found*})\ (100)}{(\text{g sample})} = \%*$$

*Of characteristic being determined.

EXAMPLE:

$$\frac{(95 \text{ g through U.S. No. 40 sieve})\ (100)}{(100 \text{ g pepper})} = 95\% \text{ through U.S. No. 40 sieve}$$

PERCENTAGE OF WEIGHT USING COMBINED DILUTIONS* AND MILLIEQUIVALENT WEIGHT:

$$\frac{(\text{ml})\ (N)\ (\text{meq wt})\ (100)\ (\text{volume*})}{(\text{g sample})\ (\text{aliquot*})} = \%$$

*May be expanded; *see* "DILUTIONS IN CALCULATIONS"

EXAMPLE:

$$\frac{(15.60\ \text{ml NaOH})\ (0.1000\ N)\ (0.06404\ \text{meq wt citric acid})\ (200\ \text{ml vol})\ (100)}{(10.00\ \text{g synthetic beverage powder})\ (10.00\ \text{ml aliquot})} = 20.0\%\ \text{citric acid}$$

PERCENTAGE USING MILLIEQUIVALENT WEIGHT IN TITRATIONS:

$$\frac{(\text{ml})\ (N)\ (\text{meq wt})\ (100)}{(\text{g sample})} = \%$$

EXAMPLE:

$$\frac{(25.20\ \text{ml NaOH})\ (0.1000\ N)\ (0.06005\ \text{meq wt acetic acid})\ (100)}{(3.000\ \text{g vinegar})} = 5.0\%\ \text{acetic acid}$$

PEROXIDE VALUE OF FATS:

$$\frac{(\text{ml})\ (N)\ (1000)}{(\text{g sample})} = \text{milliequivalents of peroxide per kilogram}$$

EXAMPLE:

$$\frac{(5.00\ \text{ml}\ Na_2S_2O_3)\ (0.100\ N)\ (1000)}{(5.000\ \text{g shortening, AOM})} = 100\ \text{meq peroxide per kg}$$

PHOSPHORUS PENTOXIDE, PERCENT:

$$\frac{(\text{ml NaOH})\ (N)\ (0.003086*)\ (100)}{(\text{g sample})} = \text{percent}\ P_2O_5$$

*Milliequivalent weight of P_2O_5 using volumetric ammonium molybdate method

EXAMPLE:

$$\frac{(22.03\ \text{ml NaOH})\ (0.1000\ N)\ (0.003086\ \text{meq wt}\ P_2O_5)\ (100)}{(5.000\ \text{g noodles})} = 0.136\%\ P_2O_5$$

POTASSIUM IODIDE IN SALT:

$$\frac{(\text{ml}\ Na_2S_2O_3)\ (N)\ (0.02776*)\ (100)}{(\text{g sample})} = \text{percent KI}$$

*Milliequivalent weight of KI by the bromine-oxidation method.

EXAMPLE:

$$\frac{(14.00\ \text{ml}\ Na_2S_2O_3)\ (0.005\ N)\ (0.02776\ \text{meq wt KI})\ (100)}{(10.00\ \text{g sample of salt})} = 0.02\%\ \text{KI}$$

POUNDS PER GALLON FROM GRAMS:

$$\frac{(3.785^{*})\ (g/ml)\ (1000)}{(28.35)\ (16)} = \text{pounds (avoirdupois) per U.S. gallon}$$

*NOTE: There are 3.785 liters per U.S. gal., 28.35 g per oz, and 16 oz per lb avoirdupois

EXAMPLE:

$$\frac{(3.785)\ (10.00\ g/10.00\ ml^{*})\ (1000)}{(28.35)\ (16)} = 8.34 \text{ lb avoirdupois per U.S. gallon}$$

*Water in air at 3.98° C

PROPORTION:

$a/b = c/d$
$a = bc/d$
$b = ad/c$
$c = ad/b$
$d = bc/a$

EXAMPLE: Problem: If there are 2 g of fat in 10 g of beef, how many g of fat would be in 140 g of beef?

$$\frac{2 \text{ g fat}}{10 \text{ g beef}} = \frac{c \text{ g fat}}{140 \text{ g beef}}$$

$$c = \frac{(2)\ (140)}{(10)} = 28 \text{ g fat}$$

PROTEIN, PERCENT:

$$\frac{(\text{ml acid})\ (N \text{ acid})\ (0.014 \text{ meq wt nitrogen})\ (\text{factor}^{*})\ (100)}{(\text{g sample})} = \text{percent protein}$$

*Factors for converting nitrogen to protein:
3.464 for caffeine
5.55 for gelatin dessert powder, and gelatin
5.7 for wheat and wheat products; alimentary pastes, baked products of wheat
6.25 for plants, cereals other than wheat, fruits, nuts, meat and meat products, dog food, wines, yeast
6.38 for milk and milk products, cheese, ice cream, etc.
20.36 for piperine
Report nitrogen in water as nitrates, and nitrogen as % *N* for fish, mayonnaise, and products not specified

EXAMPLE:

$$\frac{(32.00 \text{ ml HCl})\ (0.1000\ N)\ (0.014)\ (6.25)\ (100)}{(2.000 \text{ g sausage})} = 14.0\% \text{ protein in sausage}$$

PROXIMATE ANALYSIS: (Rational interpretation of quantitative analysis so as to approach 100% analysis)

See "CARBOHYDRATE, BY DIFFERENCE"

SALT (NaCl) PERCENT:

$$\frac{(\text{ml } AgNO_3)\ (N\ AgNO_3)\ (0.05845 \text{ meq wt NaCl})\ (100)}{(\text{g sample})} = \%\ \text{NaCl}$$

EXAMPLE:

$$\frac{(12.83 \text{ ml } AgNO_3)\ (0.1000\ N)\ (0.05845 \text{ meq wt NaCl})\ (100)}{(3.000 \text{ g meat product})} = 2.5\%\ \text{salt (NaCl)}$$

SIZE, PERCENT FOR SIEVE TESTS:

$$\frac{(\text{g passing through, or retained})\ (100)}{(\text{g sample})} = \%\ \text{through (or retained)}$$

EXAMPLE:

$$\frac{(95 \text{ g through U.S. No. 40 sieve})\ (100)}{(100 \text{ g pepper})} = 95\%\ \text{through U.S. No. 40 sieve}$$

SOLIDS IN BREAD, PERCENT:

$$\frac{(B)\ (C)}{(A)} = \%\ \text{solids}$$

A = loaf weight; B = weight after air drying; C = percent solids after oven drying

EXAMPLE: A bread roll weighed 100 g at time of receipt. The weight after air drying was 76.47 g. Percent solids after oven drying was 85%.

$$\frac{(76.47)\ (85)}{(100)} = 65\%\ \text{solids}$$

SOLIDS, PERCENT (Direct Method):

$$\frac{(\text{g solids})\ (100)}{(\text{g sample})} = \text{percent solids}$$

EXAMPLE:

$$\frac{(0.825 \text{ g solids})\ (100)}{(5.000 \text{ g mustard})} = 16.5\%\ \text{solids}$$

SOLIDS, PERCENT (Indirect Method):

$$(100\%) - (\%\ \text{moisture}) = \%\ \text{solids}$$

EXAMPLE:

$$(100\%) - (6.0\%\ \text{moisture}) = 94.0\%\ \text{solids; potatoes, white, dehydrated}$$

SUGAR IN WATER RATIO (Condensed Milk):

$$\frac{(\%\ \text{sugar})\ (100)}{(100 - \text{TMS})} = \text{ratio of sugar in water}$$

TMS = Total Milk Solids = (fat + solids-nonfat) Solids-nonfat may be abbreviated SNF

EXAMPLE:

Analytical data:

	%	
Moisture	= 28.1	(100% - % Total Solids)
Fat	= 8.9	
SNF	= 18.0	
Sucrose	= 45.0	

$$\text{Ratio} = \frac{(45.0\%)\ (100)}{(100 - 26.9)} = 61.6\%$$

$$(\text{TMS}) = (8.9\% + 18.0\%) = (26.9)$$

SULFUR DIOXIDE, PARTS PER MILLION (ppm):

$$\frac{(\text{ml NaOH})\ (N\ \text{NaOH})\ (0.032\ \text{meq wt}\ SO_2)\ (10^6)}{(\text{g sample})} = \text{ppm}\ SO_2$$

EXAMPLE:

$$\frac{(25.00\ \text{ml NaOH})\ (0.0100\ N)\ (0.032\ \text{meq wt}\ SO_2)\ (10^6)}{(32.00\ \text{g dehydrated potatoes})} = 250\ \text{ppm}\ SO_2$$

TIN PLATE (BENDIX METHOD):

(ml $Na_2S_2O_3$ of Blank - ml $Na_2S_2O_3$ for back titration) (N $Na_2S_2O_3$) (0.05935 meq wt stannous Sn) (17.28*) = pounds Sn per base box.

*Factor is based on use of sample area of 8 sq in. compared to area of 67,720 sq in. in a base box. Analysis by Bendix Method, Ind. Eng. Chem. Anal. Ed., *15*, 501 (1943).

EXAMPLE: Both sides of a disc having a total of 8 sq in. of tin plated surface were tested. (20.00 ml $Na_2S_2O_3$ - 7.80 ml $Na_2S_2O_3$) (0.1000 N) (0.05935) (17.28) = 1.25 lb tin per base box

VOLATILE OIL, ml/100 g (BY STEAM DISTILLATION):

$$\frac{(\text{ml oil})\ (100)}{(\text{g sample})} = \text{ml/100 g}$$

EXAMPLE:

$$\frac{(2.0\ \text{ml oil})\ (100)}{(100\ \text{g black pepper})} = 2.0\ \text{ml/100 g of black pepper}$$

VOLUMETRIC EQUATION, GENERAL:

$$N_1 V_1 = N_2 V_2$$

N = normality; V = volume. Any one symbol may be unknown value.

EXAMPLE:

10.00 ml of 0.1000 N HCl = V_2 of 0.0100 N HCl

$$V_2 = (10)\ (0.1000)/(0.0100) = 100.0\ \text{ml}$$